Training Circular
No. 7-100.3

TC 7-100.3
Headquarters
Department of the Army
Washington, DC, 17 January 2014

Irregular Opposing Forces

Contents

		Page
	PREFACE	iv
	INTRODUCTION	v
	Irregular Forces	v
	Irregular OPFOR and Hybrid Threat for Training	vi
	Tailoring the OPFOR	vii
Chapter 1	**IRREGULAR OPPOSING FORCE FUNDAMENTALS**	**1-1**
	Hybrid Threat	1-1
	Capabilities and Intent	1-1
	Tactics	1-4
	Interactions of Operational Variables	1-4
	Principles	1-7
	Motivations	1-10
	Support	1-12
	Blurring of Categories	1-14
	Comparison and Contrast	1-15
Chapter 2	**INSURGENTS**	**2-1**
	General Characteristics	2-1
	Insurgent Organizations	2-3
Chapter 3	**GUERRILLAS**	**3-1**
	General Characteristics	3-1
	Guerrilla Organizations	3-7
	Guerrilla Brigade	3-8
	Guerrilla Battalion	3-16
	Guerrilla Company	3-23
	Guerrilla Platoon	3-25
	Guerrilla Hunter-Killer Company	3-25
	Hunter-Killer Group	3-27
	Tactics and Techniques	3-28
Chapter 4	**CRIMINALS**	**4-1**
	Characteristics	4-1
	Motivations	4-3
	Criminal Organizations	4-4

Distribution Restriction: Approved for public release; distribution is unlimited.

Contents

	Command and Control	4-6
	Criminal Activities	4-8
Chapter 5	**NONCOMBATANTS**	**5-1**
	General Characteristics	5-1
	Armed Noncombatants	5-3
	Unarmed Noncombatants	5-7
	Unarmed Combatants	5-16
Chapter 6	**TERRORISM**	**6-1**
	Terrorism in Complex Operational Environments	6-1
	Motivations	6-2
	Terrorism Planning and Action Cycle	6-3
	Terrorism Actions	6-7
	Tactics, Techniques, and Procedures	6-9
	Terrorism Trends and Emergent Vectors	6-21
	The Specter of Terrorism	6-26
Chapter 7	**FUNCTIONAL TACTICS**	**7-1**
	Functional Organization of Forces and Elements	7-1
	Types of Offensive Action	7-3
	Types of Defensive Action	7-24
	Irregular OPFOR in Hybrid Threat Tactics	7-36
Appendix A	**INFORMATION WARFARE**	**A-1**
	General Characteristics	A-1
	INFOWAR Organizations	A-4
	INFOWAR Support to Irregular OPFOR	A-6

Figures

Figure I-1. Irregular force actors v
Figure 1-1. Relation of irregular OPFOR principles to achieving the objective 1-8
Figure 2-1. Network types 2-5
Figure 2-2. Typical levels of leadership, action, and support 2-6
Figure 2-3. Higher insurgent organization 2-10
Figure 2-4. Local insurgent organization 2-13
Figure 2-5. Local and higher insurgent organizations network (example) 2-14
Figure 2-6. Insurgent direct action cell and team graphic symbols 2-15
Figure 2-7. Insurgent supporting cell graphic symbols 2-20
Figure 3-1. Guerrilla organization symbols: brigade to team level 3-7
Figure 3-2. Guerrilla brigade (example) 3-8
Figure 3-3. Guerrilla battalion (example) 3-17
Figure 3-4. Guerrilla company (example) 3-24
Figure 3-5. Guerrilla platoon (example) 3-25
Figure 3-6. Guerrilla hunter-killer company (example) 3-26

Figure 5-1. Armed and unarmed noncombatants (examples) 5-1
Figure 5-2. Noncombatant support to the irregular OPFOR (examples) 5-3
Figure 5-3. PSC organization by functions (example) 5-6
Figure 5-4. Media team organization (example) 5-9
Figure 5-5. NGO field office functions (example) 5-12
Figure 6-1. Terrorism actors in complex operational environments 6-1
Figure 6-2. Terrorism planning and action cycle 6-4
Figure 6-3. Bombing: insurgent affiliate (example) 6-11
Figure 6-4. Kidnapping: insurgent-criminal affiliation (example) 6-13
Figure 6-5. Hostage-taking: guerrilla unit (example) 6-14
Figure 6-6. Raid: multiple insurgent suicide cell assaults (example) 6-15
Figure 6-7. Anthrax-laced letters: death-mayhem in targeted population (example) 6-21
Figure 7-1. Insurgent ambush (example) 7-4
Figure 7-2. Guerrilla ambush (example) 7-5
Figure 7-3. Insurgent assault (example) 7-10
Figure 7-4. Guerrilla assault (example) 7-11
Figure 7-5. Insurgent raid (example) 7-11
Figure 7-6. Guerrilla raid (example) 7-11
Figure 7-7. Guerrilla reconnaissance attack (example) 7-11
Figure 7-8. Simple and complex battle positions 7-25
Figure 7-9. Insurgent defense of a simple battle position (example) 7-28
Figure 7-10. Guerrilla defense of a simple battle position (example) 7-29
Figure 7-11. Insurgent defense of a complex battle position (example) 7-33
Figure 7-12. Guerrilla defense of a complex battle position (example) 7-34

Tables

Table 1-1. Comparison and contrast of insurgents, guerrillas, and criminals 1-16
Table A-1. INFOWAR elements, objectives and targets A-3

Preface

This training circular (TC) is one of a series that describes an opposing force (OPFOR) for training U.S. Army commanders, staffs, and units. See the References section for a list of other TCs in this series. (Other publications in the former Field Manual [FM] 7-100 series will be converted to TCs as well.) Together, these TCs outline OPFOR that can cover the entire spectrum capabilities of regular and/or irregular forces against which the Army must train to ensure success in any future conflict.

Applications for this series of TCs include field training, training simulations, and classroom instruction throughout the Army. All Army training and red team venues should use an OPFOR based on these TCs, except when mission rehearsal or contingency training requires maximum fidelity to a specific country-based threat or enemy. Even in the latter case, trainers should use appropriate parts of the OPFOR TCs to fill information gaps in a manner consistent with what they do know about a specific threat or enemy.

This publication applies to the Active Army, the Army National Guard (ARNG) /Army National Guard of the United States (ARNGUS), and the United States Army Reserve (USAR) unless otherwise stated.

Headquarters, United States Army Training and Doctrine Command (TRADOC) is the proponent for this publication. The preparing agency is the Complex Operational Environment and Threat Integration Directorate (CTID), TRADOC G-2 Intelligence Support Activity (TRISA)-Threats. Send comments and suggested improvements on DA Form 2028 (Recommended Changes to Publications and Blank Forms) directly to CTID at the following address: Director, CTID, TRISA-Threats, ATTN: ATIN-T (Bldg 467, Rm 15), 803 Harrison Blvd, Fort Leavenworth, KS 66027-1323.

This publication is available at on the General Dennis J. Reimer Training and Doctrine Digital Library (RDL) at http://www.adtdl.army.mil. Also, see Army Training Network (ATN) https://atn.army.mil with Army Knowledge Online (AKO) access. Two locations on the ATN front-page access TRADOC G-2 products. Click the "Training for Operations" button and then click the "CTID Operational Environment Page" link. Click the "DA Training Environment" button and click the "OPFOR & Threat Doctrine" link. Both of these sites provide threat and opposing forces (OPFOR) resources and references for training, professional education, and leader development. Updates subject to the Army Publishing Directorate (APD) approval process will occur as required, and as a result of a normal review production cycle for an Army training circular. The date on the cover and title page of the electronic version will reflect the latest change update.

Introduction

This training circular (TC), as part of the TC 7-100 series, addresses the irregular opposing force (OPFOR), which in Army training exercises represents a composite of actual threats and enemies that comprise irregular forces. The three primary categories of irregular forces portrayed by the OPFOR are insurgents, guerrillas, and criminals. These actors may operate separately or in conjunction with one another and/or combined with regular military forces as the Hybrid Threat (HT) for training. This TC also addresses other actors present in an operational environment (OE), who may be affiliated with the irregular OPFOR through willing support or coercion, and/or may be passive or unknowing supporters of the irregular OPFOR.

IRREGULAR FORCES

Irregular forces are armed individuals or groups who are not members of the regular armed forces, police, or other internal security forces (JP 3-24). The distinction of being armed as an individual or group can include a wide range of people who can be categorized correctly or incorrectly as irregular forces. Excluding members of regular armed forces, police, or internal security forces from being considered irregular forces may appear to add some clarity. However, such exclusion is inappropriate when a soldier of a regular armed force, policeman, or internal security force member is concurrently operating in support of insurgent, guerrilla, or criminal activities.

Irregular forces can be insurgent, guerrilla, or criminal organizations or any combination thereof. Any of those forces can be affiliated with mercenaries, corrupt governing authority officials, compromised commercial and public entities, active or covert supporters, and willing or coerced members of a populace. Independent actors can also act on agendas separate from those of irregular forces. Figure I-1 depicts various actors that may be part of or associated with irregular forces.

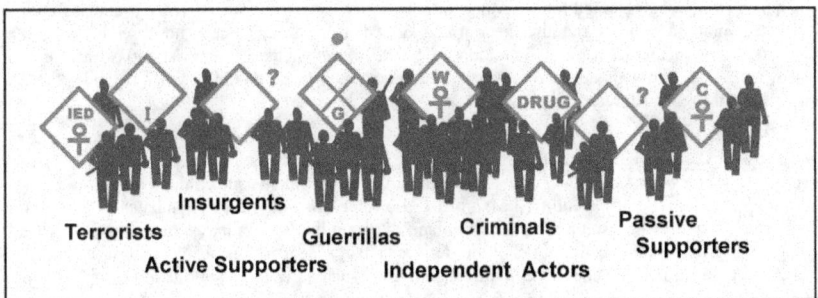

Figure I-1. Irregular force actors

Closely related to the subject of irregular forces is irregular warfare. JP 1 defines *irregular warfare* as "a violent struggle among state and non-state actors for legitimacy and influence over the relevant population(s). Irregular warfare favors indirect and asymmetric approaches, though it may employ the full range of military and other capacities, in order to erode an adversary's power, influence, and will." The definition spotlights a dilemma of conflict in and among a population. It also indicates that the non-state actors characterized as irregular forces may operate in other than military or even military-like (paramilitary) capacities.

Introduction

IRREGULAR OPFOR AND HYBRID THREAT FOR TRAINING

The function of the OPFOR is to portray a threat or enemy in training exercises. Army Regulation (AR) 350-2 establishes policies and procedures for the Army's Opposing Force Program. Since 2004, that AR has defined *opposing force* as "a plausible and flexible military and/or paramilitary force representing a composite of varying capabilities of actual worldwide forces used in lieu of a specific threat force, for training or developing U.S. forces." AR 350-2 has defined *threat* as "any specific foreign nation or organization with intentions and military capabilities that suggest it could become an adversary or challenge the national security interests of the United States or its allies." However, that definition of *threat* was much narrower than the way ADRP 3-0 now defines it.

"A *threat* **is any combination of actors, entities, or forces that have the capability and intent to harm U.S. forces, U.S. national interests, or the homeland.** Threats may include individuals, groups of individuals (organized or not organized), paramilitary or military forces, nation-states, or national alliances. When threats execute their capability to do harm to the United States, they become enemies." (ADRP 3-0)

A *hybrid threat* **is the diverse and dynamic combination of regular forces, irregular forces, terrorist forces, and/or criminal elements unified to achieve mutually benefitting effects** (ADRP 3-0). A hybrid threat can consist of any combination of two or more of those components.

The preface of TC 7-100 says that TC outlines "the Hybrid Threat that represents a composite of actual threat forces as an opposing force (OPFOR) for training exercises." After focusing on real-world hybrid threats in its first part, TC 7-100 states: "Part two of this TC focuses on the Hybrid Threat (HT) for U.S. Army training. The HT is a realistic and relevant composite of actual hybrid threats. This composite constitutes the enemy, adversary, or threat...represented as an opposing force (OPFOR) in training exercises." That means that, in the context of the HT for training, the OPFOR can also include nonmilitary actors (such as criminal elements), which are part of the threat faced by the training unit. Since the definition of *hybrid threat* includes other than military and paramilitary actors, it has **broadened the application of the term OPFOR**.

The AR 350-2 definition of OPFOR used the term *paramilitary forces*, defined in JP 3-24 as "forces or groups distinct from the regular armed forces of any country, but resembling them in organization, equipment, training, or mission." However, the ADRP 3-0 (and TC 7-100) definition of *hybrid threat* uses the **broader** term *irregular forces*, defined in JP 3-24 as "armed individuals or groups who are not members of the regular armed forces, police, or other internal security forces." Thus, irregular forces would include paramilitary forces, but also other armed individuals or groups that do not resemble regular armed forces. Although the ADRP 3-0 definition of *hybrid threat* includes "terrorist forces, and/or criminal elements," both criminals and actors who use terrorism as a tactic would actually fit under the JP 3-24 definition of irregular forces.

FM 7-100.4 lists criminal organizations under non-state paramilitary actors, but only because "Criminal organizations at the higher end of the scale can take on the characteristics of a paramilitary organization." However, it notes that "individual drug dealers and criminals or small-scale criminal organizations (gangs)" do not have that type of capability. FM 7-100.4 also refers to "other armed combatants" as "nonmilitary personnel who are armed but not part of an organized paramilitary or military structure. Nevertheless, they may be disgruntled and hostile." ARDP 3-0 defines "a party identified as hostile against which the use of force is authorized" as an enemy. In a training exercise, such an enemy would be portrayed as part of the OPFOR.

FM 7-100.4 also recognizes the category of "unarmed combatants." It says: "The local populace contains various types of unarmed nonmilitary personnel who, given the right conditions, may decide to purposely and materially support hostilities against the United States. ... Individual criminals or small gangs might be affiliated with a paramilitary organization and perform support functions that do not involve weapons. ... Even unarmed individuals who are coerced into performing or supporting hostile actions and those who do so unwittingly can in some cases be categorized as combatants. Thus, various types of unarmed combatants can be part of the OPFOR."

Thus, a threat or enemy can be any individual or group—**not necessarily military and/or paramilitary, or even armed**. The same is true of the OPFOR that represents the threat or enemy in a training exercise.

TAILORING THE OPFOR

The OPFOR must be a challenging, uncooperative adversary or enemy. It must be capable of stressing any or all warfighting functions and mission-essential tasks of the U.S. armed force being trained. This TC addresses the irregular OPFOR with a tactical focus on insurgents, guerrillas, and/or criminals. In exercise design (see TC 7-101), the type(s) of forces making up the OPFOR will depend upon the conditions determined to be appropriate for accomplishing training objectives. In some cases, the OPFOR may only need to reflect the nature and capabilities of a regular military force, an irregular force, or a criminal organization. However, in order to be representative of the types of threats the Army is likely to encounter in actual OEs, the OPFOR will often need to represent the capabilities of a hybrid threat. In that context, the force that constitutes the enemy, adversary, or threat for an exercise is called *the Hybrid Threat*, with the acronym HT. Whenever the acronym is used, readers should understand that as referring to the Hybrid Threat. The HT is a realistic and representative composite of actual hybrid threats.

Regular military forces of the OPFOR may have irregular forces and/or criminal elements acting in loose affiliation with them, or acting separately from them within the same training environment. These relationships depend on the scenario, which is crafted based on the training requirements and conditions of the Army unit being trained. The OPFOR tactics described in TC 7-100.2 are appropriate for use by an OPFOR that consists either entirely or partly of regular military forces. Some of those tactics, particularly those carried out by smaller organizations, can also be used by insurgents, guerrillas, or even criminals. Even those tactics carried out primarily by regular military forces may involve other components of the HT acting in some capacity. When either acting alone or in concert with other components of the HT, irregular forces and/or criminal elements can also use other tactics which are outlined in TC 7-100.3.

> *Note.* After this introduction, the chapters and appendixes of this TC address their topics from the OPFOR point of view. So, *friendly* refers to the irregular OPFOR and other parts of the Hybrid Threat. Likewise, *enemy* refers to the enemy of the irregular OPFOR, which may be an opponent within its own country or region or an extraregional opponent (normally the United States or a U.S.-led coalition).

This page intentionally left blank.

Chapter 1
Irregular Opposing Force Fundamentals

This chapter describes the fundamental characteristics of the irregular opposing force (OPFOR). This TC portrays three main categories of irregular OPFOR: insurgents (chapter 2); guerrillas (chapter 3); and criminals (chapter 4). Other actors present in an operational environment (OE) can be willing, coerced, and/or unknowing participants in irregular OPFOR activities (chapter 5). The irregular OPFOR interacts with the eight operational variables that make up an OE. This interaction among variables adds to the complexity of an OE and creates multiple opportunities for the irregular OPFOR to optimize its capabilities against an enemy. The irregular OPFOR can use terrorism as a tactic (chapter 6) in addition to military-like functional techniques (chapter 7). A table in this chapter compares and contrasts the capabilities and limitations characteristic of a guerrilla force, an insurgent organization, and a criminal organization. The particular capabilities and limitations of the various types of irregular OPFOR are described in subsequent chapters of this TC.

HYBRID THREAT

1-1. The irregular OPFOR can be part of the Hybrid Threat (HT). (For more information of the HT, see TC 7-100.) The HT can be any combination of two or more of the following components: regular military forces, irregular forces, and/or criminal elements. The irregular OPFOR component of the HT can be insurgents, guerrillas, or criminals or any combination thereof. The irregular OPFOR can also include other armed individuals or groups who are not members of a governing authority's domestic law enforcement organizations or other internal security forces. On occasion, situations may occur where unarmed individuals or groups may be part of the irregular OPFOR and the HT. An example of unarmed individuals aligned with the HT in an active support role is when segments of the population participate in public demonstrations against an enemy of the HT. Possible HT combinations include parts of the irregular OPFOR operating openly with regular military forces, being sponsored directly or indirectly by a state's government, or being supported by non-state organizations.

1-2. The irregular OPFOR can be part of the HT, but can also operate independently without any allegiance to or collaboration with other types of forces associated with the HT. Various state and non-state organizations, regular military forces, paramilitary forces, and/or criminal organizations might be operating in the same space and time as the irregular OPFOR but not be part of the irregular OPFOR or the HT. To be a hybrid threat, all these components would have to be "unified to achieve mutually benefitting effects" (ADRP 3-0 and TC 7-100).

CAPABILITIES AND INTENT

1-3. The irregular OPFOR is adaptive, flexible, and agile. It can quickly change its composition to optimize organizational capabilities and use those capabilities against known or perceived vulnerabilities of an enemy. The irregular OPFOR takes prudent risks when an expectation exists for successful attack on an enemy. However, it may also make significant practical sacrifices in individuals and materiel in order to achieve a major psychological impact on an enemy. An example of such deliberate sacrifice is a number of nearly simultaneous, small unit or direct action cell assaults on targets that result in the deaths of all attackers, but receive sensational media coverage to a global audience.

1-4. The intent of the irregular OPFOR is to acquire a range of capabilities and use them at selected times and locations in order to achieve desired effects. It can use those capabilities against an enemy. However, the irregular OPFOR can also use functional tactics and/or terrorism to manipulate a population and dissuade support to an enemy's military forces and/or other institutions. When necessary, it will use acts of violence to gain influence and develop willing or coerced cooperation. Concurrently, it will use indirect means to progressively degrade an enemy's physical power and infrastructure, and to psychologically influence the political, social, economic, military, and information variables of an OE. The irregular OPFOR will attempt to exploit its familiarity with the physical environment and its ability to blend into the local populace. The time variable normally favors the irregular OPFOR. The activities of the irregular OPFOR are constant over a long period of time, but may change in pace, tempo, and speed. The timing of actions will sometimes appear random when the actual mode of the irregular OPFOR and its activities are deliberate decisions as part of a long-term campaign or strategy.

1-5. One of the most significant capabilities of the irregular OPFOR is the ability to manipulate and/or ignore the restrictions and sanctions that apply to regulated military forces, law enforcement agencies, and internal security forces belonging to a sovereign state. International protocols and conventions, national statutes and law, and moral codes that guide behavioral norms and social interactions can limit the enemy's use of weapon systems and other capabilities that overmatch irregular OPFOR capabilities. The irregular OPFOR can make exceptions by complying with these codes of conduct when that is advantageous for its information warfare campaign. However, it can easily abandon those standards when they no longer provide operational value. When regular military forces of a state incorporate clandestine use of the irregular OPFOR, the state can often plausibly deny responsibility for actions conducted by irregular forces.

1-6. Although violent actions by any individual organization or combination intend to receive immediate notoriety, the irregular OPFOR complements violent actions with methodical, long-term psychological warfare. The overarching agenda of the irregular OPFOR can include but is not limited to the following issues:
- Expand support and/or control within an area or region.
- Deter opposition to its objectives within a relevant population.
- Obtain popular recognition and support of its objectives by designated segments of a population.
- Marginalize the governance and/or extraregional influence of an adversary.
- Attract an international or global audience and/or external sources of influence to support irregular OPFOR aims.

1-7. The irregular OPFOR seeks to gain the approval and support of at least certain elements of a relevant population in order to obtain active or passive assistance. The methods by which it acquires such influence are complex in any OE. Normally, it must communicate a compelling narrative of legitimacy that is accepted by the population. This credential of legitimacy may require a gradual process of convincing the relevant population that the irregular OPFOR is an acceptable means to achieve desired social, religious, or political effects. However, the irregular OPFOR may confer authority on itself without regard to the population's goals.

1-8. An enemy of the irregular OPFOR may maintain that the OPFOR concept of legitimacy is corrupt and illicit. The irregular OPFOR may declare that its actions are justifiable under existing conditions, and attempt to degrade the legitimacy of a governing authority. Over time, the irregular OPFOR seeks to obtain recognition of its legitimacy by a willing populace and official recognition from external states and/or organizations in order to accomplish its long-term goals. Once its authority is recognized, the irregular OPFOR seeks to maintain the legitimacy of its cause, its leadership role, and its actions.

1-9. Sometimes external recognition and support is not as important to the irregular OPFOR as is establishing a geographic enclave from which to plan, prepare, and conduct its activities and influence. The irregular OPFOR conducts direct and indirect actions that are adaptive and persistent. This form of conflict incorporates irregular forces typically categorized as insurgents or guerrillas, and includes selective actions coordinated with criminal organizations. Particular actions can be purposely conducted as acts of terrorism, or can employ more military-like tactics. All of these actions are described in terms of the common functional framework described in the 7-100 series of FMs and TCs.

COMPLEXITY AND COLLABORATION

1-10. The irregular OPFOR may be part of a complex array of irregular and regular OPFOR organizations, units, or individuals with various coordinated and/or disparate single-agenda aims. A particular geographic, political, cyberspace, or ideological environment may lead to alliances or affiliations that are dynamic and constantly changing. Discrete incidents may not seem to be part of an overall plan. However, detailed analysis of the irregular OPFOR actions and associated political, social, economic, information, and other events normally reveals a vision supporting a long-range aim.

1-11. In particular conditions and circumstances, irregular OPFOR actions can include support from regular military forces and/or special-purpose forces (SPF) from a state or states. Internal security forces and/or law enforcement organizations that have been infiltrated by the irregular OPFOR can also support irregular OPFOR actions within an area or region. The collaboration among organizations, units, and/or individuals of a relevant population may be based on coercion, contractual agreement, and/or temporary or long-term common goals and objectives.

1-12. Possible rationales for irregular OPFOR collaboration with other organizations or individuals in an OE can include—

- Spotlight grievances for resolution.
- Establish influence over a relevant segment of a population.
- Develop acceptance and legitimacy of irregular OPFOR programs and actions.
- Achieve OPFOR objectives without alienating critical segments of indigenous and/or extraregional populations.
- Cultivate active or passive supporters.

1-13. Irregular OPFOR objectives may promote solutions to grievances in the context of a particular population. The irregular OPFOR may prefer to use indirect approaches such as subterfuge, deception, and nonlethal action to achieve its objectives. However, it is committed to violent action, when necessary, in order to compel an enemy and/or an opposing form of governance to submit to its intentions. Some irregular OPFOR organizations such as affiliated criminal gangs exist for their own commercial profit and power, and are not interested in the quality of life and/or civil security of a relevant population that they influence or coerce. Other forms of the irregular OPFOR can be rogue individuals with single-issue agendas who are willing to use criminal activity and/or terrorism in order to achieve their objective.

ADAPTABILITY

1-14. The irregular OPFOR is constantly adapting its capabilities in an agile and flexible manner to achieve its objectives. These capabilities include improvements in organization, equipment, and tactics. The irregular OPFOR can readily task-organize for a particular action. It tailors actions to support a compelling agenda that resonates with a relevant population for active and/or passive support. It makes adjustments when it gains or loses affiliated support or experiences degradation due to recent actions.

1-15. Irregular OPFOR actions are conducted as a continuum. Any pause in its operations is part of a coherent campaign of persistent conflict. A long-term perspective guides near- and mid-term actions to marshal capabilities for future actions. While one form of action may appear stalled, another form of action is likely underway against an enemy weakness. Protracted actions can change quickly if the irregular OPFOR observes unexpected enemy vulnerabilities.

1-16. The irregular OPFOR's ability to quickly transition also gives it the agility and flexibility to—

- Command, control, and/or influence various activities.
- Task-organize its forces.
- Deceive and surprise.
- Disperse and concentrate.
- Retain freedom of movement.
- Apply physical and psychological techniques in order to create anxiety in an enemy.

This agility and flexibility is critical to how effectively the irregular OPFOR adapts its patterns of operations to maintain the initiative over an enemy. The irregular OPFOR perseveres in adversity by its ability to adapt.

1-17. FM 7-100.4 (to be converted to a TC) outlines the baseline organizations and equipment of insurgents, guerrillas, and criminals and provides general guidance on how these forces may task-organize for particular missions. This process of task-organizing and/or tailoring of forces is flexible to accommodate the particular mission and conditions in the OE.

> *Note.* In order to achieve a desired level of capabilities, the irregular OPFOR order of battle or task-organized structure may require adjustments in types and numbers of weapons systems, other equipment, and personnel. If a particular piece of equipment in the organizational directories of the OPFOR administrative force structure in FM 7-100.4 is not appropriate for a specific OE, possible substitutions are identified in the *Worldwide Equipment Guide*.

TACTICS

1-18. Insurgents and guerrillas, as part of the irregular OPFOR, may employ variants of the functional tactics outlined for smaller tactical units of the regular OPFOR in TC 7-100.2 (see chapter 7 of this TC). Criminal elements do not normally have the ability to execute these functional tactics.

1-19. All three types of forces that make up the irregular OPFOR can employ terrorism as a tactic in order to achieve their aims. (For actions typical of criminals, see chapter 4. For more detail on terrorism, see chapter 6.)

INTERACTIONS OF OPERATIONAL VARIABLES

1-20. The irregular OPFOR is part of the military variable, which explores the military and/or paramilitary capabilities of all relevant actors (enemy, friendly, and neutral) in a given OE. However, irregular OPFOR actions can affect or be affected by all the operational variables: political, military, economic, social, information, infrastructure, physical environment, and time (PMESII-PT) and their subvariables. The impacts may be robust, moderate, or relatively insignificant. The interaction of the operational variables and subvariables establishes conditions for various levels of irregular OPFOR capabilities and limitations. The dynamic interaction and effects by the irregular OPFOR on operational variables are a physical and psychological combat multiplier for the OPFOR. The following paragraphs discuss the impacts of the irregular OPFOR on each of the operational variables. They also provide examples of how the actions in one variable can directly or indirectly impact on other variables and affect the capabilities and influence of the irregular OPFOR.

POLITICAL

1-21. The centers of responsibility and power at various levels of governance can be an objective for irregular OPFOR subversion or violent action. The irregular OPFOR may target for willing or coerced support of its aims—
- Constituted authorities at local, provincial, and/or state levels.
- Tribal leaders and/or clan chiefs.
- Religious leaders and councils.
- Influential political organizations.

The irregular OPFOR may want to institute its own political goals for the perceived benefit of a relevant population, or to create chaos within a governing authority in order create a protected geographic enclave within a sovereign state. In either case, influence over a relevant population is essential. Given enough electoral support in a relevant population, the irregular OPFOR may be able to win formal political recognition at varied levels of a governing authority that it opposes.

1-22. Conditions of the political variable would very likely interact with social and economic conditions of an OE. The political aim of the irregular OPFOR could have a genuine intent to provide a voice in politics to an under-represented relevant population, or be self-serving in order to obtain control of political institutions for its own commercial profit. It may start as the former and transition into the latter. An example of this transition could be if the irregular OPFOR emerged from an indigenous population of rural farmers and tenants with grievances against absentee landowners and corporate businesses. However, over time, the irregular OPFOR might shift its focus to commercial profit in racketeering and the production, and distribution of illegal drugs to a transnational market.

MILITARY

1-23. The irregular OPFOR can infiltrate regular military and/or paramilitary forces of an enemy governing authority. It can collect intelligence on military unit strengths and weaknesses, unit leader preferences and biases, and/or readiness of weapons, support, and materiel. Covert actions by irregular OPFOR members can undermine the effectiveness of enemy units by—

- Raising doubts about the validity of enemy unit missions.
- Subverting leader and subordinate allegiances.
- Questioning the general treatment of military and/or paramilitary members by the governing authority.

1-24. With military training, weapons and equipment stolen or purchased by the irregular OPFOR can be used more effectively to improve its armed capabilities. In some cases, the irregular OPFOR may attempt to transition highly trained irregular forces into a more formalized paramilitary or security force in order to demonstrate its ability.

1-25. An example of interaction of military and information variables could be a media campaign directed at a local, regional, and transnational audience, in which the irregular OPFOR uses the symbols and appearance commonly associated with military power and influence. The irregular OPFOR could use progressive success in establishing and protecting a geographic safe haven with a declaration of sovereignty. Announcements to a global information network could display senior irregular OPFOR leaders in military uniform and attire, speaking publicly with official banners or flags, and maps or images of territory declared as independent. A focused defense of the safe haven, assisted by an extensive and supportive diaspora could achieve irregular OPFOR objectives unless challenged by an enemy governing authority.

ECONOMIC

1-26. The economic effectiveness and prosperity of a governing authority can be marginalized with black market activities coordinated by the irregular OPFOR. Criminal activities can include smuggling, theft, and/or piracy of marketable goods. Insurgent or guerrilla actions can include disrupting the flow of commerce throughout the economic chain of production, distribution, and consumption by the general populace. The irregular OPFOR, in some instances, can become the illicit commercial broker for what transactions occur in an economic sector. Front companies and/or organizations can launder resources and money into legitimate enterprises in support of irregular OPFOR objectives.

1-27. An example of interaction between economic, infrastructure, and information variables could be when the irregular OPFOR can deliver a satisfactory level of livelihood, health care, and commercial advancement for a relevant population. The population may have been denied access to such expectations by transnational corporations and a governing authority that extract natural resources from the region for their own profit and exclusive use. The grievance of this economic poverty can result in irregular OPFOR actions such as sabotage of pipelines, disruption of refining facilities, kidnapping of corporate officials, and/or random acts of murder. The irregular OPFOR can use a media campaign aligned with the economic grievances of the relevant population to emphasize the validity of OPFOR offensive actions. For example, it could call attention to the presence of significant private security contractors of transnational corporations and regular military forces of a governing authority conducting business security actions and military operations for their own financial gain.

SOCIAL

1-28. Cultural, religious, and ethnic differences can be stress points within a population that the irregular OPFOR can incite with real or false claims to further fracture a society and its social institutions. Unsettled grievances based on traditional values and customs can range from dissatisfaction to violent demonstrations against issues such as human rights, educational opportunities, and/or social mobility. The irregular OPFOR can nurture the support of particular social and religious leaders of a community that align themselves with OPFOR initiatives. Civic improvement associations and social welfare projects administered by the irregular OPFOR focused on a relevant population can be part of a comprehensive social unity program.

1-29. An example of interaction of the social and information variables could be the compelling influence that a cleric or advisory council of clerics can have on a relevant population. Religious leaders may have traditional authority in a culture that recognizes such an overarching authority. Based on that perceived authority, the directives and prohibitions of that cleric or council can direct popular support of the irregular OPFOR or civil disobedience to a governing political authority.

INFORMATION

1-30. Public communications by the irregular OPFOR can promote its own message with perception management and also discredit programs of the enemy governing authority. Open source and human intelligence collection operations can improve the irregular OPFOR awareness of enemy weaknesses. Information warfare (INFOWAR) by the irregular OPFOR can deceive, destroy, or sabotage enemy communication infrastructure. Examples of irregular OPFOR INFOWAR activities include blogs on the Internet, publicizing incidents of alleged abuse by a governing authority in global news media, and/or recurring presentations by charismatic individuals in social, political, educational, and religious institutions.

1-31. Another example of how the interaction of variables can cause significant impacts on irregular OPFOR operations is the way the physical environment can influence the transmission and flow of irregular OPFOR information and intelligence. In a rural setting, the irregular OPFOR may use normal commercial transportation means of porters and mule trains. In an urban setting, it may use vendors and daily routes of commercial goods. Thus, in either setting, sophisticated electronic monitoring capabilities by an enemy of the irregular OPFOR may become insignificant. The irregular OPFOR can manipulate the global information environment by using the Internet to encrypt and/or hide messages in innocent appearing electronic communications. Verbal or electronic communications of a relevant population can be stifled by irregular OPFOR threats and/or coercion when such information negatively affects its operations in an area. Public claims of responsibility for acts of terror by the irregular OPFOR can further reduce overt support of a relevant population to an enemy of the irregular OPFOR.

Note. The information variable is discussed in more detail in appendix A, which describes the elements of the irregular OPFOR's INFOWAR.

INFRASTRUCTURE

1-32. The irregular OPFOR can cause immediate, progressive, and long-term effects on of basic facilities, services, and operational installations that provide the infrastructure for a community or society. Immediate impacts can be the disruption of a sole-source oil pipeline and/or the destruction of a main electrical power generation facility. Progressive acts to disrupt and cause long-term impacts can be the mining of regional motor transportation routes with improvised explosive devices and recurring ambushes of commercial and military convoys. From a more positive perspective, the irregular OPFOR could provide basic civil services not being provided by a governing authority such as potable water and/or basic medical care.

1-33. Irregular OPFOR actions can include calculated disruption or destruction activities to limit enemy capabilities and/or convince a relevant population that the enemy is incapable of protecting them and providing adequate public or private services. The irregular OPFOR may choose to—
- Attack public utilities or the civil servants who service such utilities.
- Prevent delivery of commodities that sustain performance of public services.

- Interdict services such as motor transportation, urban construction and repair.
- Disrupt implementation of farming initiatives hosted by an enemy governing authority.

PHYSICAL ENVIRONMENT

1-34. The irregular OPFOR can optimize the geography and/or manmade structures of a physical environment by using restrictive rural terrain or dense urban communities to mask complex battle positions and safe havens. Terrain, weather, climate, and vegetation can support the concealment and/or cover of irregular OPFOR staging and conduct of operations. The growth of poor and underemployed populations in many large metropolitan areas, and the increase of some cities to a status of megalopolis are prime conditions to create grievances that can be manipulated by the irregular OPFOR.

1-35. An example of interaction among variables could be the impact of a governing authority's land reforms that adversely affect the immediate prosperity of a local or regional population. This could include a directed destruction of one type of farming crop on limited usable terrain, and a replacement crop that does not provide the same economic value to the farmer or local distributor. This situation can be part of INFOWAR, with the irregular OPFOR promoting a story of how the governing authority it to blame for these negative impacts.

TIME

1-36. Time can be a combat multiplier for the irregular OPFOR when the cultural perception of time accepts a protracted conflict. During such a protracted conflict, the irregular OPFOR can use violent actions, INFOWAR, diplomatic discussions, economic pressures, and progressive representations of value-added for a relevant population. Timing and duration of activities, events, or conditions, as well as how the timing and duration are perceived by various actors in the OE, can prevent or delay governing authority activities in favor of irregular OPFOR aims.

1-37. The irregular OPFOR seeks to chose the time and place for engaging the enemy. Timing can be the most significant aspect of determining when to tactically execute a decision to delay, deceive, fix, and/or block. Timeliness of information and intelligence is another key aspect that the irregular OPFOR uses to its own advantage. In order to affect enemy pace, tempo, and/or speed of action and reaction, the irregular OPFOR may plant false information at a particular time and ensure that an enemy obtains it. The prudent use of time is often combined with characteristics of a physical environment to create opportunities in support of near-, mid-, and/or long-term objectives.

PRINCIPLES

1-38. The fundamental principles of the irregular OPFOR guide organizational and individual actions. Regardless of whether the organization is military-like in appearance and operations, or includes elements that are loosely affiliated with the irregular OPFOR in a political, social, or economic sector of a society, these principles provide a common framework of how the irregular OPFOR plans and conducts actions. Since the irregular OPFOR often resembles military forces in many ways, some of these principles are very similar to fundamental military principles in the regular OPFOR. In some cases, however, the principles are tailored in consideration of irregular OPFOR capabilities or limitations. Depending on the type of irregular OPFOR organization and its goals and motivations, the irregular OPFOR adapts the following principles to address particular situations. Figure 1-1 on page 1-8 shows how each of the principles interrelate and contribute toward achieving an irregular OPFOR objective. That objective typically relates to a particular grievance on the part of the irregular OPFOR or of a relevant population it seeks to influence.

INITIATIVE

1-39. Initiative is the ability of the irregular OPFOR to retain a freedom of action in its plans and operations. Initiative enables the irregular OPFOR to force an enemy to react to its actions. Success often goes to the side that conducts itself more actively and resolutely. Irregular OPFOR leaders encourage initiative to make and implement bold decisions in order to establish or change the terms of the irregular conflict in favor of the irregular OPFOR. Subordinates are expected to take advantage of new developments immediately. They seek to overcome a position of relative inferiority while operating within a senior OPFOR leader's intentions. Initiative exploits an enemy's restrictive rules of engagement or political restrictions.

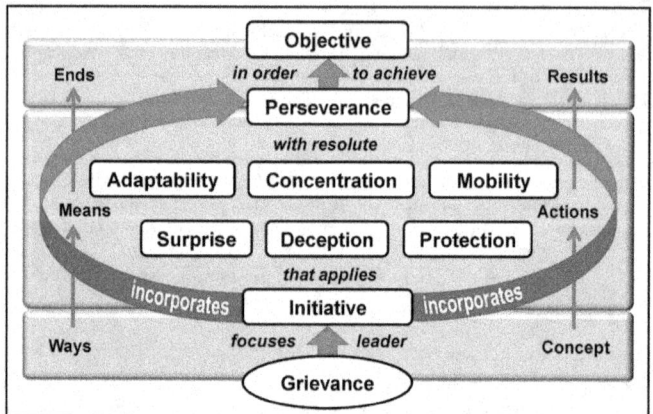

Figure 1-1. Relation of irregular OPFOR principles to achieving the objective

DECEPTION

1-40. Deception is the ability to deliberately convey a false and/or distorted picture of the situation to an enemy leader that is targeted for deception. Deceptive information causes an enemy leader to believe he has accurate situational awareness and understanding. Irregular OPFOR leaders plan and direct deception that helps them accomplish their objective, but does not hamper other concurrent OPFOR actions. Feints and demonstrations are examples of deception. Other elements of INFOWAR attempt to optimize the effects of deception in tactics and techniques. Various irregular OPFOR capabilities and actions can lead to a compelling yet inaccurate analysis by the enemy leader. These means of deception, not limited by domestic or international law and conventions, can include the following:
- Physical activity or lack of activity.
- Technical resources and employment techniques indicative of a particular tactic.
- Administrative indicators such as visual, oral, or documentary evidence in order to manipulate and distort reality.

SURPRISE

1-41. Surprise is the ability of the irregular OPFOR to take advantage of an enemy vulnerability in a manner for which an enemy is unprepared or unable to effectively counter. Irregular OPFOR action is normally swift and fleeting and may employ unexpected means. The irregular OPFOR attempts to shape a setting so that an enemy is not expecting the action or create conditions that an enemy is not prepared to confront. The irregular OPFOR achieves surprise through deception activities in conjunction with protection and security measures and/or other elements of INFOWAR. Surprise can be achieved by means such as the following:
- Changing tactics, techniques, or the intensity of actions against an enemy.
- Employing commercial or industrial materiel as a weapon in unexpected ways.
- Presenting public indications of compromise or cessation of armed conflict. For example, the irregular OPFOR may agree to a truce and series of negotiations with an enemy, but use the time to prepare and stage for nearly simultaneous surprise attacks on military and civil government facilities.

PROTECTION

1-42. Protection is the ability to preserve irregular OPFOR effectiveness of its organizational assets and capabilities. These assets and capabilities include OPFOR personnel, equipment, weapon systems, operations, information, facilities, and/or infrastructure. Protection involves a continuous, integrated series and/or group of measures that sustain the ability for the irregular OPFOR to plan, prepare, and conduct successful actions. Protection and security measures are a key element in INFOWAR (see appendix A). The irregular OPFOR normally operates with a minimal or unidentifiable signature within a relevant population in order to avoid being a lucrative target. Protection complements the use of a safe haven, when required, to refit or reconstitute irregular OPFOR combat power. An example of protection is an irregular OPFOR guerrilla unit exfiltrating from a raid in hostile territory to a secure location in a neighboring state. This unofficial support by a state near the area of OPFOR actions provides security and protection to the guerrilla unit while it recruits and trains replacements due to casualties incurred in the raid.

MOBILITY

1-43. Mobility is the ability to sustain irregular OPFOR freedom of movement within areas controlled or occupied by the enemy. The irregular OPFOR seeks to create an advantage over the enemy regarding knowledge and use of geographic terrain and populations in order to position, reposition, and/or prepare for and conduct effective actions. A high degree of mobility enables the irregular OPFOR to use available combat power with maximum effect at a decisive time and place. For example, the irregular OPFOR can blend into a population with similar clothing and daily habits in order to maintain anonymity while transiting an area or region. This type of mobility can allow the timely supply of weapons and materiel, fiscal resources, and/or manpower for designated actions in an area or region.

ADAPTABILITY

1-44. Adaptability is the ability of the irregular OPFOR to use initiative and creative thinking in order to set particular conditions and take advantage of the resulting opportunities. Irregular OPFOR leaders recognize emergent developments that change existing conditions, and apply initiative that causes the enemy to react at a disadvantage to the actions of the irregular OPFOR. Simple tactics and techniques can be adapted for use against an enemy's sophisticated technology and weapon systems. For example, a system of couriers can negate the intrusive ability of electronic monitoring devices to detect and locate the financial transaction networks of a local insurgent organization that were formerly conducted with cellular telephones. Conversely, the irregular OPFOR can adapt to sophisticated techniques such as encrypting and hiding information within harmless appearing communiqués such as electronic files, images, and documents transiting the Internet.

CONCENTRATION

1-45. Concentration is the ability of the irregular OPFOR to mass the capabilities of combat power in time and space, in order to achieve a desired effect. Concentration of effort allows the irregular OPFOR to create and dominate a condition for a specific amount of time. When the irregular OPFOR concentrates rapidly or gradually from dispersed locations to conduct a particular action, a normal subsequent action is to quickly disperse in order to avoid an effective enemy response against the massed OPFOR. An example of concentration is the coordination to quickly mass irregular OPFOR insurgent cells, and use a swarming technique in a raid to overwhelm a designated target at an isolated combat outpost. Once insurgents have seized the objective and secured or destroyed designated weapons, equipment, and documents, the cells quickly disperse into the countryside or urban areas in order to avoid capture or death.

PERSEVERANCE

1-46. Perseverance demonstrates the will of the irregular OPFOR to persist in long-term commitment to fight an enemy until it accomplishes its goals and objective. Protracted and persistent operations are the norm of the irregular OPFOR. Actions may be subtle and can be part of a gradual series of actions toward achieving a task. Dramatic individual actions are often used to establish or sustain notoriety for the irregular OPFOR, but rarely achieve a decisive effect. Periodic setbacks in irregular OPFOR missions are anticipated and rationalized with effective INFOWAR announcements to sustain a moral dominance of the irregular OPFOR over an enemy. INFOWAR techniques can also convince a relevant population to sustain support of the irregular OPFOR even though its actions may require an extended period of time before the population eventually benefits. The irregular OPFOR may attempt to achieve its objectives within a specified timetable and announced milestones; however, the objectives may take decades or generations to achieve.

MOTIVATIONS

1-47. Insurgents and guerrillas are normally motivated by social, religious, or political issues or some combination those. In most cases, criminals—whether they are part of the irregular OPFOR or not—have other motivations. Although individual actors who conduct criminal activity may not be part of the irregular OPFOR, the individual's personal motivation and/or ideology may be the deciding perspective of *why* he or she acts.

> *Note.* The local populace may provide active or passive support out of a different motivation than the insurgents or guerrillas they support. For example, the motivation of the populace might be financial (payment or beneficial effects on business profits) or security provided them. The populace might provide support, based on ethnic or religious issues, to an insurgent or guerrilla organization even if they do not share that organization's political agenda.

INSURGENTS OR GUERRILLAS

1-48. Certain types of motivation are common to insurgents and guerrillas, the two most likely components of the irregular OPFOR. However, insurgents and guerrillas that agree to collaborate against a common enemy may or may not share the same motivations or ideology.

> *Note.* If multiple insurgent and/or guerrilla organizations exist in a particular OE, they may share some motivations but differ in others. In order to form an affiliated relationship, the organizations just need to have one or more motivations that coincide with or complement each other. An example of a coinciding motivation could be that both organizations resent the presence of an extraregional force in their country. An example of a complementary relationship would be if one organization has financial resources, while the other needs financial support

1-49. The motivation that incites violent as well as nonviolent actions by the irregular OPFOR is often framed in the context of ideology. The irregular OPFOR acts in a particular way based on underlying grievances, which are often linked to the ideals of an ideology. These unresolved grievances—perceived or factual—create conditions where armed and unarmed individuals believe they must act to obtain what they believe is a just solution. The rationale and the resulting actions may be perceived in a positive and negative light by a relevant population. The motivation and rationale typically are one or more of the following:

- Personal or group social identity.
- Devotion to a particular religious belief.
- Commitment to a form of political governance.

Combinations among these factors can further complicate how to describe the motivation and ideology of a particular person or group. Other motivations exist and may become a primary prompt for action, though only loosely associated with a social, religious, or political agenda. Aspects of ethnicity, geography, and history affect personal and group relationships as witnessed in social status and networks, religion, and politics. Combinations can occur among these factors to further complicate how to describe the motivation of a particular person or group.

Social Identity

1-50. Social identity, as an individual and/or as a member of a social group, is often a fundamental aspect of why people may be aligned with or alienated from the irregular OPFOR. They may be attracted by an ability to satisfy a perceived critical want or need in their lifestyle. Allegiance to a clan, tribe, or familial grouping is an example of social identity and accountability. These forms of social allegiance can indicate why change is desired or required in a social order. The same rationales can support why a set of ideals or practices must be sustained or expanded within a relevant population. Several common categories of social identity that can overlap with religious or political agendas are as follows:

- Ethnocentric groups who understand race or ethnicity as the defining characteristic of a society and basis of cohesion.
- Nationalistic groups who promote cultural-national consciousness and perhaps establishment of a separate nation-state.
- Revolutionary groups who are dedicated to the overthrow of a governing authority and establishment of a new social order.
- Separatist groups who demand independence from an existing governance that appears socially, theologically, or politically unjust to a relevant population.

1-51. A variation on these identity categories is an independent actor who conceives, plans, and conducts violent or nonviolent actions without any direction from another person or irregular force. This type of individual may be sympathetic to the aims of a particular group, but have no contact with the group or affiliated members of the group.

Religion

1-52. Religion can be a compelling motivation. As a personal belief, religion can be interpreted as divine edict and infallible. Practice of a faith system is a personal interpretation and decision. However, fundamentalist clerics and/or religious mentors in some instances can interpret passages of religious doctrine in a particular way that supports the irregular OPFOR agenda. Some interpretations may be a purposeful misrepresentation never intended by an original religious author. Other clerics or radical splinter groups may honestly believe in a religious duty to pursue a fundamentalist approach to worship and lifestyle. In either case, religion can be used as a catalyst to instigate rivalry between and among ethnic and/or religious denominational groups. Confrontations can be provoked by faith system practices that are completely unacceptable to another faith group and/or can create an irreconcilable theological wedge between cultural-faith groups in a relevant population.

1-53. The irregular OPFOR may attempt to link religion with declarations that governing political authorities who do not accept a particular understanding of a faith doctrine are a wicked secular presence that must be destroyed and replaced with a fundamentalist theocracy form of governance. Cults, although not a religion by normal definition, can adopt similar forms of violence, mass murder, and mayhem as part of a self-proclaimed apocalyptic vision and purpose.

Politics

1-54. The irregular OPFOR, especially the insurgent part, can have a political agenda. Prevalent political systems that can be promoted by the irregular OPFOR include forms of governance such as—

- Single-party totalitarian state.
- Nationalist-fascist authority.
- Leadership appointed or democratically elected by popular vote.
- Social self-management and equality aimed at reducing or eliminating political and economic hierarchies.
- Social-political process that evolves toward a classless-stateless society with common ownership on means of production, communal access to commodities for livelihood, and general social programs for the benefit of community and human wellness.
- Pseudo-social and political authority that protects a geographic sanctuary and/or safe haven for conduct of an illicit commercial enterprise.

1-55. The irregular OPFOR can use motivations of a political ideology to attract the attention of a relevant population in order to develop influence with a particular community. This can lead to support and/or collaboration or actually joining the irregular OPFOR. Since politics are integral to the overarching conditions that affect daily life, commerce, vocation, freedoms, and family, the irregular OPFOR can use political influence—along with social and/or religious appeal—to enhance its legitimacy. Visible actions, often localized in perspective, focus on demonstrating power and authority.

CRIMINALS

1-56. Although actors who conduct criminal activities may not be insurgents or guerrillas, personal motivation and/or ideology may be a deciding factor in their decision to join the irregular OPFOR voluntarily or be coerced to participate. Criminals who become part of the irregular OPFOR may or may not share some of the same motivations as other parts of the OPFOR. Regardless of their social identity and religious or political preferences, criminals are motivated primarily by money. (See chapter 4 for more detail on the motivations of criminals and criminal organizations.)

SUPPORT

1-57. Support for irregular OPFOR activities includes capabilities provided by individuals and organizations internal and external to the geographic area. Multiple forms of support can be voluntary and/or coerced from individuals, selected segments of a population, or activities and organizations with particular capabilities. Affiliation with and/or among organizations indicates a common goal. However, affiliation is not a formal association and subordination among participants. Support relationships are often temporary and remain in effect only as long as the benefits of cooperation are mutual to the involved organizations.

INTERNAL

1-58. Internal support includes the ideological and practical means to improve and/or sustain irregular OPFOR operations. Insurgents or guerrillas seek to ally and affiliate with activities and people that range from specific commodities and/or functional skills to networks of political, commercial, religious, and/or social groups with required capabilities. Internal support indicates the commitment of leaders, active supporters, and passive supporters to further the progress of the irregular OPFOR. As conditions change in geographic, political, cultural, cyberspace, and/or ideological environments, the coordination of internal support adjusts to accommodate emergent opportunities.

EXTERNAL

1-59. External support is not always necessary for the irregular OPFOR to succeed. However, some form of external support is usually critical to sustain long term operations. Activities and organizations often provide external support for ideological reasons, but other motivations can include political power, commercial enterprise, financial wealth, and/or some common grievance affecting a relevant population. The motivations that incite the irregular OPFOR affect how and why support is obtained. Whether support is voluntary or coerced, the rationale of why support is necessary reflects how discrete aspects of assistance improve a cumulative capability.

1-60. Prominent categories of support are moral, money, and materiel. The morality of supporting the irregular OPFOR links an ideology to an outcome of desired conditions by a relevant population. Money enables the irregular OPFOR to acquire capabilities required in order to enhance its influence not only in a paramilitary sense, but also across a full spectrum of operational variables in an OE. Materiel support indicates the provision of equipment, supplies, and operating systems used by the irregular OPFOR and/or its supporters. Materiel requires training, maintenance, and recurring systems improvement. These capabilities suggest the need for trainers, advisors, and/or liaison personnel to support the irregular OPFOR. To accomplish these actions, a safe haven or sanctuary allows a freedom of action in planning, preparing, and staging for overt and covert operations. Three main types of external support are diaspora support, state support, and non-state support.

Diaspora

1-61. A diaspora is a relevant population that has voluntarily moved or migrated, and/or been forcefully displaced from an established or ancestral homeland. However, they still have racial, national, tribal, religious, linguistic, and/or cultural affinities with inhabitants of that homeland. Motives for support from a diaspora can include ethnocentric, separatist, and/or nationalistic. A fundamental objective is to correct proven or perceived injustices to a group of people in a larger community and relevant population. This type of support is often linked to a common social, religious, and/or political identity.

1-62. Voluntary support from a diaspora is a preferred method for assistance to the irregular OPFOR. However, compelling the support from segments of a diaspora may be required at times to ensure the amount of general or specialized support required to sustain operations. Compelling support from a diaspora is a calculated risk that is closely monitored by irregular OPFOR senior leaders in order to not alienate a significant portion of the relevant population.

1-63. Beyond aspects such as financial and materiel contributions, the expansion of support through INFOWAR with and for a diaspora can sway public opinion and obtain continuing extraregional and international support for the struggle of the irregular OPFOR. In the spotlight of globalized media, a diaspora usually accents the moral correctness of resolving proven or perceived injustices to a group of people in a relevant population of the particular geographic region or state.

External State

1-64. Support from an external state for the irregular OPFOR can be covert and/or overt. A state acts primarily based on its political objectives. Political, economic, social, fiscal, moral, and/or logistics support can be provided while disavowing any direct connection to the irregular OPFOR. For example, a state may condemn an adversary state for repressing a relevant population while covertly providing arms, ammunition, sanctuary, training bases, and/or advise and other forms of assistance.

1-65. A state might employ SPF and/or covert agents of the state to conduct direct actions in support of the irregular OPFOR. The state officially denies responsibility or state sponsorship for such activities, but can engage in unofficial INFOWAR releases that indicate support for specified actions within the insurgency. In other situations, a state may openly declare a legal or moral right to intervene in the affairs of another state and support designated insurgent organizations and their affiliates. States may form a coalition to condemn the acts of a governing authority and announce support for irregular OPFOR actions.

1-66. A state can act overtly by providing supplies, training, and other forms of support to the irregular OPFOR. This support can be coordinated and provided without any specified control over an insurgent or guerrilla organization by the state. Other options can exist where stipulations are mutually agreed upon between representatives of the state and irregular OPFOR leaders in exchange for materiel and/or a safe haven or protected staging area for operations. Other means of support can include—

- Access to training facilities and expertise not otherwise readily available to the irregular OPFOR.
- Extension of diplomatic protections and services such as immunity from extradition.
- Use of embassies and other protected grounds.
- Use of diplomatic pouches to transport weapons or explosives.
- False documentation for personal identification and movements throughout geographic regions and states.

1-67. States in a coalition or as members of an international organization may decide to intervene directly in an insurgency with diplomatic envoys, observation teams, and/or military forces. A coalition may form to condemn the acts of a governing authority and announce support for irregular OPFOR actions. The irregular OPFOR insurgent organization leaders and coalition authorities can mutually agree to—

- Seek conditions for justice and respect arising from international treaties and other sources of international law.
- Promote social progress, economic advancement of indigenous people.
- Maintain international stability and security.

Non-State

1-68. Forms of non-state support can range from non-state transnational networks with global reach capability to individuals with a special or unique capability. These forms of support can be overt and/or covert. Networks can be regional or international in their aims, and/or individuals committed to a single-issue agenda. Whether support involves a type of network, hierarchical organization, or loosely coordinated individuals, non-state support to an insurgency or independent guerrilla operations often includes sophisticated criminal activities in foreign and/or indigenous environments.

1-69. Insurgencies may turn to transnational criminal organizations for funding. Other globalized capabilities can provide advanced knowledge and technology exchanges, mobile international transportation, and near-instantaneous cyber communication. Non-state support exchanges in capabilities can occur directly and indirectly.

1-70. Transnational economic entities such as international business corporations may provide support to irregular OPFOR operations in order to further their corporate business interests. Other supervisory or workforce individuals in non-profit or private organizations may assist the irregular OPFOR based on personal beliefs and/or as the result of coercion or extortion. Individual financiers may provide support in order to advance personal agendas or cultural, ethnic, political, and/or religious causes.

1-71. External support from other parts of the irregular OPFOR and/or in conjunction with regular military forces can include an exchange of personnel and equipment, as well as intelligence, training, recruitment, logistics, and finances. In addition to individual and contracted mercenaries, a significant body of other independent or loosely collaborative individuals may be available for hire to an irregular OPFOR organization. These personnel may have served previously in guerrilla units or regular military units.

BLURRING OF CATEGORIES

1-72. Although three basic types of forces can be part of the irregular OPFOR, the distinctions among insurgents, guerrillas, and criminals are sometimes blurred. That is because they may have more in common than they have that is different. From the viewpoint of the existing government authority, for instance, the activities of all three types are illegal, that is, criminal. Not just criminals but also insurgents and guerrillas can engage in criminal activities. Some insurgent organizations can include guerrilla units (developed from within or affiliated) and some guerrilla units may be part of an insurgency. In advanced phases of an insurgency, guerrilla units may begin to look and act more like regular military units.

1-73. There are three general tactics available to the irregular OPFOR—
- Military-like functional tactics.
- Criminal activity.
- Terrorism.

1-74. At any given time, the irregular OPFOR could use any of these means. The differences among these three can become blurred, especially within an urban environment or where the governing authority exerts strong control.

INSURGENTS AND GUERRILLAS

1-75. As an insurgent organization grows, its cellular structure or parts thereof may develop into a more hierarchical military-like structure. Within an insurgent organization, direct action cells may combine to form guerrilla squads and platoons. In some cases, a platoon or company may retain some direct action cells (possibly for terror tactics). Other specialized cells of an insurgent organization can evolve into parts of—
- Supporting units of a guerrilla company, battalion, or brigade.
- Staff elements of a guerrilla battalion or brigade.

1-76. Guerrillas may engage in more military-like operations after an insurgency develops extensive popular and logistics support. When they feel the conditions are set, insurgents may generate a conventional military force that can directly confront regular forces of the existing governing authority.

INSURGENTS AND CRIMINALS

1-77. There is often a nexus between insurgency and crime. Sustainment requirements, especially funding, often drive insurgents into relationships with organized crime or into criminal activity themselves. Insurgents may turn to local, regional, or transnational criminal elements for funding. Cooperating with criminals may not be ideologically consistent with the movement's core beliefs, although that does not necessarily prevent such cooperation.

1-78. An insurgent organization itself may engage in criminal activities as a source of funding. Income is essential not only for insurgents to purchase weapons but also to pay recruits and bribe corrupt officials. Reaping windfall profits and avoiding the costs and difficulties involved in securing external support makes illegal activity attractive to insurgents. Kidnapping, extortion, robbery, and trafficking (drug, human, black market goods, and so on)—four favorite insurgent activities—are very lucrative, although they can also alienate the population.

1-79. Insurgents may use coercive force to gain power over the population. Examples of organizations providing such force are insurgent direct action cells, guerrilla units, gangs, and organized crime elements. Such groups may use their coercive means for a variety of purposes unrelated to the insurgency. Protecting their community members, carrying out vendettas, and engaging in criminal activities are examples. Insurgent organizations may also attract followers through criminal activities that provide income.

1-80. Insurgencies attract criminals and mercenaries. Fighters who joined for money will probably become bandits once the fighting ends and may engage in criminal activity during the fighting. This category includes opportunists who exploit the absence of governing authority security and law enforcement to engage in economically lucrative criminal activity, such as kidnapping, smuggling, or theft.

1-81. Some insurgencies can compartmentalize criminal activity, keeping it ancillary to the main effort and preventing it from affecting the organization and its unity. However, some insurgencies can become focused on criminal activity that once only served as a funding mechanism. Some insurgent cell leaders may become crime bosses. The insurgency as a whole may degenerate into criminality, particularly if the primary movement disintegrates and the remaining elements are cast adrift. Such disintegration replaces an ideologically inspired body of individuals with a more diverse body, normally of very uneven character.

1-82. Even when criminal networks are not a part of an insurgency, their activities—for example, banditry, hijackings, kidnappings, and smuggling—can further undermine the governing authority. Insurgent organizations often link themselves to criminal networks to obtain funding and logistics support. In some cases, insurgent networks and criminal networks become indistinguishable. Most insurgent groups are more similar to organized crime in their organizational structure and relations with the populace than they are to military units.

REGULAR AND IRREGULAR OPFOR

1-83. There can also be some blurring between the irregular OPFOR and regular OPFOR military units. Both types of forces can use many of the same functional tactics (such as assault, ambush, and raid).

1-84. The regular OPFOR, especially SPF, may use terror tactics similar to those of the irregular OPFOR. Since SPF may or may not be in uniform, they may be hard to distinguish from irregular OPFOR insurgents or guerrillas, for which they often serve as trainers or advisors and alongside which they may fight.

COMPARISON AND CONTRAST

1-85. Within the irregular OPFOR, guerrilla units, insurgent organizations, and criminal organizations have various capabilities and limitations. Table 1-1 on pages 1-16 through 1-18 compares and contrasts the basic characteristics of the three types of forces in order to highlight their similarities and their differences.

Table 1-1. Comparison and contrast of insurgents, guerrillas, and criminals

Characteristic	Insurgent Organization	Guerrilla Unit	Criminal Organization
Leadership	Network is typical, but can include hierarchical sub-organizations; leaders may be located distant from the geographic area of conflict; political or ideological mentors or council advisors and/or counsel senior leaders. Title for personnel in command is usually "leader." Some leaders may use a religious, historical, or honorific title.	Hierarchical with military-like chain of command and control or support systems; leaders predominantly indigenous; political advisors may accompany guerrilla units in actions. Leader titles are military in nature such as battalion and/or company commander, platoon leader, section leader, team leader, hunter-killer group leader.	Hierarchical structure or network dependent on origin of organization; even in small criminal organizations, leaders may be located distant from the geographic area of conflict; political or ideological mentors or council advise and/or counsel senior leaders. Leader titles can be traditional or historical terms, or simple authority terms.
Motivation	Insurgency movement with a political and/or ideological agenda. Can also be social identity or religion.	Social identity, religion, or politics. Can be military component to an insurgency; or can be independent of an insurgency with a specified agenda.	Intention to profit fiscally through control of a process, commodity, and/or area; social identity as a power broker in a designated geographic, economic, or social environment.
Organization	Cellular-network model; can be hierarchical for designated capabilities or functions; can include paramilitary capability for a primarily political-oriented organization; can be affiliated with other irregular OPFOR and/or regular military forces.	Military unit model with echelons of command and control; can include land, sea, and air capabilities; can be affiliated with other irregular OPFOR and/or regular military forces; more likely that other irregular OPFOR components to be closely integrated with regular military forces.	Hierarchical structure or network dependent on origin of organization; general categories of gangs, large-scale syndicates, and transnational organizations; organizations can be based on family, ethnic, commodity, or specialized purpose. Can infiltrate or become affiliated with insurgent, guerrilla, or regular military forces.
Objectives	Concessions from and/or defeat of a political opponent; ultimately, overthrow an enemy governing authority and replace governance with insurgent movement leadership; seek legitimacy as movement.	Military mission success within a campaign in support of unit goals and desired end state; can be an independent and specified guerrilla unit agenda; can be the military capability in insurgent organization.	Profit from activities and coercion; expand organizational influence within an area, regional, or transnational scope; preserve control of specified commodities, geographic areas, and/or services; avoid contact with governing authority.
Internal Support	Active and passive support in local and larger area population; can have legitimate social-economic-political activities to mobilize civil support.	Active and passive support by segments of a local area population for military-type capabilities; can expand support to regional area population.	Active and passive support in local and larger area population; can use coercion to influence legitimate social-economic-political activities or individual support.
External Support	Regional safe havens; Diaspora systems to promote insurgent movement in regional and/or international communities; can receive cooperation or assistance from regular forces, SPF, or state activities opposing the governing authority.	Regional safe havens; can receive cooperation or assistance from regular military, SPF, or state activities opposing the governing authority in the area of guerrilla operations.	Cooperative affiliations among gangs, large-scale syndicates, and/or transnational organizations can provide designated support and services; co-opted governing authority offices may also assist.
Activity Patterns	Local, regional, provincial, and/or district activities with intention of obtaining support of relevant population; can be social, economic, diplomatic, political, and military activities.	Military-like functional tactics as norm; can expand tactical actions into a military campaign focused in a geographic area.	Local, regional, and/or transnational activities; random or systematic activity to sustain influence; specialized expertise can be part of functional business model of larger commercial enterprises.

Table 1-1. Comparison and contrast of insurgents, guerrillas, and criminals (continued)

Characteristic	Insurgent Organization	Guerrilla Unit	Criminal Organization
Composition	Higher insurgent organization and local insurgent organization; functional structure with cellular base; can receive assistance or support from SPF and/or regular military forces.	Brigade to squad-team guerrilla units; can be tailored for specialized missions; can be supported by SPF advisors and/or regular forces advisors or liaison.	Transnational organization, large-scale syndicates, and/or gangs; can be groups affiliated in temporary or long-term specific business arrangements.
Personnel	Indigenous and/or transnational core; ideological mentor support network; rank and file members can be in a period of active service in the insurgent organization, or be authorized inactive status and return to familial-social responsibilities with their local population.	Predominantly indigenous core; can include members outside of indigenous population; rank and file members typically remain in the same military unit with a long-term commitment of duty.	Predominantly indigenous core, but can include members outside of indigenous population as organization expands in business connections; status may indicate familial or social lineage within organization; rank and file members may be specialized operators.
Recruitment	Local, regional, or extraregional population recruiting; often relates to an ideological or faith system commitment to serve; can be contract support; can use conscription for limited lengths of time or seasonal terms of service.	Local or regional population recruiting base; informal induction process with expectation of long-term or open-ended term of service; can use conscription from local population; require compliance with strict standards and discipline.	Familial or social affiliations may be membership requirement; rank and file members may be contracted for skill sets; initiation rites and tests may precede acceptance; code of conduct may demand absolute allegiance to the organization.
Weapons and Other Equipment	Tier 1-4 capabilities (see *Worldwide Equipment Guide*) dependent on support from insurgent region and diaspora; can receive clandestine support from states, private organizations, and/or criminal activities that oppose the governing authority in conflict with the insurgent organization.	Tier 1-4 capabilities (see *Worldwide Equipment Guide*) dependent on access to local resources, or a higher and/or local insurgent organization if part of an insurgency; if independent, obtains weapons from raids, ambushes, and black market purchases; can include sophisticated systems and heavy weapon capabilities similar to regular military forces.	Tier 1-4 capabilities (see *Worldwide Equipment Guide*) dependent on ability to purchase or procure weapons and/or materiel.
Uniforms	Local attire of populace as norm; can wear military-like pieces of clothing and load-bearing equipment for designated direct actions.	Varied military-like clothing or pieces of clothing, load-bearing equipment, and insignia as norm; can wear local attire of populace to facilitate mission.	Attire of local populace as norm; can use civilian or law enforcement or paramilitary pieces of clothing and equipment in order to deceive.
Disposition	Within political boundary of governing authority as norm; can be safe haven in or near contested territory; size and location of cells based on local conditions. Rural and/or urban areas on mission basis; often occupy safe havens in populace, can locate as dispersed cells, and/or in CBP; can use social, economic, or political organizations as public and visible presence in populace.	Within enemy-held, hostile, or denied territory; can be safe haven in or near geographic area of contested territory; unit size, location, and command integrity based on local conditions. Rural and/or urban areas on mission basis; often occupy safe haven and complex battle position (CBP) away from populace; can locate units in proximity to other guerrilla units for mutual support.	Within geographic area under organization control; can be safe haven in or near area involved in business transactions; size and location of organizational groupings based on local conditions. Rural and/or urban on mission basis; often use protected locations in populace, can locate as dispersed groups; co-opted social, economic, or political organizations can prevent interference by law enforcement or internal security organizations.

Table 1-1. Comparison and contrast of insurgents, guerrillas, and criminals (continued)

Characteristic	Insurgent Organization	Guerrilla Unit	Criminal Organization
Training	Functional skills in local or regional safe havens, or contract support for niche expertise; skills and services can include administrative, intelligence, direct actions, supply, transportation, communications, and/or special skill sets; can be augmented by SPF and/or regular military advisors.	Basic and advanced military skills instructed in safe haven or CBP in or near geographic area of conflict. If part of insurgency, coordinates for special skills and support; can be augmented by SPF and/or regular military advisors.	Basic and advanced skills learned in practice; special skills and support can be contracted; training can include administrative, intelligence, internal security, operational franchise management, supply and services, transportation, communications, and/or special skill sets.
Internal Security	Counterintelligence and internal security cells monitor insurgent loyalty.	Counterintelligence elements monitor guerrilla loyalty.	Internal security maintained through codes of conduct and monitoring activities and outside contacts; specialized groups monitor member loyalty.
Logistics	Self-sustaining with populace support; support from regional safe havens and diaspora; can receive regular military and SPF materiel support; can commandeer materiel from local area.	Self-sustaining with populace support; can receive mission-based support from regular military forces and SPF; can barter and/or commandeer materiel from local area; can capture supplies and materiel from enemy forces.	Self-sustaining with supply, services, and materiel purchased, stolen, or coerced from a local populace, or provided by a regional or transnational network; can be supported from regional safe havens.
Communications	Local, regional, and/or global; easy access to media; access to Internet and INFOWAR technologies; secure and encryption capable; common use of messenger or courier.	Local and regional; tactical communications norm; secure and encryption capable; access to Internet and INFOWAR technologies; common use of messenger or courier.	Local, regional, and/or global; easy access to media; access to Internet and INFOWAR technologies; secure and encryption capable; common use of messenger or courier.
Finances	Periodic reward in goods and services; may include contract-for-hire payment systems; can be obligation with no pay.	Semiformal or formal pay system in local script and/or barter materiel; can be conscription with no pay.	Payment to designated members based on performance and profit; payments may be cash, services, and or in-kind barter; investments and payments are often laundered to mask origin.
Terrorism	Use or restriction of terrorism can be due to insurgent proclamation; psychological weapon; actions are integrated as part of INFOWAR campaign; purpose ultimately supports a political objective can be focused or random for effects.	Use or restriction of terrorism based on guerrilla commander directive; psychological weapon; actions are integrated as part of INFOWAR campaign; can be deliberate or random for intended effects.	Use or restriction of terrorism can be due to senior or intermediate leader decision; field operators may use terrorism within the domain they control; psychological weapon; actions can be focused or random for effects.

Chapter 2
Insurgents

This chapter presents an overview of insurgent organizations and actions as part of the irregular OPFOR for U.S. Army training. The insurgent irregular OPFOR is representative of threats in a resistance movement and/or insurgency that can exist in various operational environments (OEs). Insurgents can be armed or unarmed. In addition to functional tactics (see chapter 7), insurgents can use acts of terrorism (see chapter 6) to intimidate or influence a governing authority or a relevant population.

GENERAL CHARACTERISTICS

2-1. *Insurgents* are armed and/or unarmed individuals or groups who promote an agenda of subversion and violence that seeks to overthrow or force change of a governing authority. They can transition between subversion and violence dependent on specific conditions. Both types of action intend to disrupt a governing authority. They gradually undermine the confidence of a relevant population in a governing authority's ability to provide and justly administer civil law, order, and stability. Insurgents can achieve their aims without violence, but this is not the norm.

RELATIONSHIPS WITH OTHER ORGANIZATIONS AND ACTORS

2-2. During an insurgency, the distinctions among insurgents, guerrillas, criminals, and other actors in an OE are often unclear. Insurgents may use deception to add to this confusion and sometimes deny responsibility for direct actions or acts of terrorism. Insurgent organizations can act separately from other groups, organizations, and/or activities in conflict with an enemy or in conjunction with them to achieve common goals.

2-3. Insurgents may conduct operations in combination with regular military forces of a state in conflict with the governing authority that insurgents oppose. Advisors, liaison teams, and military forces—especially from special-purpose forces (SPF)—can provide overt and covert support for insurgent actions.

2-4. When present in the irregular OPFOR, guerrillas are normally incorporated into and subordinate to a higher insurgent organization. However, guerrilla forces can also be aligned with local insurgent organizations. In some cases, lower-level guerrilla units may be subordinate to a local insurgent organization. Guerrilla forces can also exist as an independent capability, completely independent of higher and local insurgent organizations. Evolving conditions may cause affiliations or task organizations that are particular to that context. The relationships between insurgents and guerrillas operating in an area may be temporary and remain in effect only as long as the both organizations mutually benefit. Guerrillas are described in detail in chapter 3.

2-5. Insurgents may also act in conjunction with criminal elements. Criminals may exploit the instability caused by insurgency to further their own profit. Insurgents may consort with criminals or resort to criminal activities themselves, in order to finance and sustain their operations.

> *Note.* The Hybrid Threat (HT) used in training of U.S. Army forces can include all these types of actors and capabilities. The HT can be any combination of two or more of the following components: regular forces, irregular forces, and/or criminal elements. Possible HT combinations include insurgents and other parts of the irregular OPFOR operating openly with regular military forces, being sponsored directly or indirectly by another state's governmental ministries and/or departments, or supported by non-state organizations. (See TC 7-100 for detailed discussion of the HT.)

2-6. Insurgent organizations can form temporary affiliations with other commercial, social, or political entities. They can also solicit or coerce the active and/or passive support of civilians in and outside of the area of conflict. Such supporters can include—
- Those merely sympathetic to the goals of the insurgents.
- Those providing monetary support.
- Those actively supporting and engaging in direct actions alongside the insurgents and/or guerrillas.

Insurgents may have the overt and/or covert support of a government-in-exile when both organizations view a governing authority as a common enemy.

SCOPE

2-7. Insurgent organizations normally conduct irregular conflict within or near the sovereign territory of a state in order to overthrow or force change in that state's governing authority. Some insurgent activities—such as influencing public opinion and acquiring resources—can occur outside of the geographic area that is the focus of the insurgency.

2-8. An insurgent organization may begin or remain at the local level. A *local insurgent organization* may exist at small city, town, village, parish, community, or neighborhood level. It may expand and/or combine with other local organizations. Cities with a large population or covering a large area may be considered regions and may include several low-level insurgent organizations. A *higher insurgent organization* may exist at regional, provincial, district, national, or transnational level. Higher insurgent organizations usually contain a mix of local insurgent and guerrilla organizations. The higher insurgent organization can apply both types of forces with a wider scope of impact. The OE and the specific goals determine the size and composition of each insurgent organization and the scope of its activities. (See sections on Higher Insurgent Organizations and Lower Insurgent Organizations, below, for more detail.)

LINES OF EFFORT

2-9. An insurgency is fundamentally a political movement. The expectation of a long-term conflict requires plans for and use of physical and psychological force. Civic actions develop, expand, and marshal the support of a relevant population for the insurgency's agenda. A comprehensive plan of action typically incorporates three main lines of effort:
- Political influence.
- Direct action violence and terrorism.
- Civic interaction and support.

Political Influence

2-10. The political element provides the overarching command and control (C2) of the insurgent organization. The political leadership plans and directs the strategy and actions to divide or weaken the governing authority they oppose. Information warfare (INFOWAR) activities foster dissatisfaction of the relevant population with the governing authority and show the insurgency as an opportunity for change. The insurgency degrades the confidence of the population in the governing authority. At the same time, the political element is preparing and/or implementing its own administrative and governance capabilities that provide solutions to the population's grievances.

Direct Action Violence and Terrorism

2-11. Insurgent cellular organization provides an adaptable function-based capability. *Direct action cells* reside primarily in local insurgent organizations and usually conduct small-scale and focused violent acts at the tactical level of conflict. (Direct action cells are described in detail later in this chapter.) Actions can range from one-person tasks to multiple cells tailored temporarily for specific operations. Subversion and selective or random violence are planned acts to incite frustration and overreaction by a governing authority. The government reaction can anger the relevant population and further undermine its allegiance or passive support to the governing authority. If an insurgency advances to the use of *guerilla units*, the guerrillas conduct operations against elements of the governing authority with functional tactics. (See chapter 3 for detail on guerrillas and chapter 7 on functional tactics.) Terrorism can be applied throughout these direct and supporting actions. (See chapter 6 on terrorism.)

2-12. Targets are often the activities and organizations that provide civil law and order such as—
- Police.
- Civil administrators.
- Internal security forces.
- Regular military forces of the governing authority.

2-13. Insurgent actions can range from simple threats and hoaxes to use of sophisticated technology and weapon systems. Insurgent organizations generally do not possess much of the heavier and more sophisticated equipment the guerrilla organizations possess. If the insurgents require these weapons or capabilities, they may either obtain them from guerrillas, or the guerrilla organization may provide its services depending on the relationship between the two organizations at the time. Some insurgent organizations profess the desire to acquire and use weapons of mass destruction (WMD).

Civic Interaction and Support

2-14. Civic interaction by insurgents with a relevant population establishes and maintains influence over the population. It allows the insurgency to successfully organize clandestine actions. Activities can—
- Incite open demonstrations against the governing authority.
- Improve recruitment efforts for the insurgency within the population.
- Demonize the governing authority as a threat to the population rather than protector.

2-15. If a relevant population believes that a governing authority is incapable of effective governance, insurgents can usually obtain increasing active and passive support from the population. Insurgents degrade the operational effectiveness of a governing authority in selected functions. They use a full range of INFOWAR capabilities to exhaust the resolve of a governing authority and increase the population's will to support the overthrow of such governance.

2-16. Insurgents conduct some civic initiatives in a manner that does not overtly link the services to it. Other actions often assist in providing basic social services that mitigate suffering of the population and provide support. The insurgents will want to take credit for these. Such support can be humanitarian programs for—
- Food.
- Potable water.
- Basic medical services and preventive medicine.
- Basic safe guarding of personal and commercial property.
- Arbitrating civil and social disputes ignored by a governing authority.

2-17. The insurgent organization can sometimes obtain enough public support its representatives can be legally elected to political positions within a governing authority. Ultimately, the insurgency must convince an increasing number of uncommitted citizens and passive supporters within the relevant population to accept its agenda for change or replacement of the current governing authority.

INSURGENT ORGANIZATIONS

2-18. Insurgent organizations do not have a fixed structure. The mission, environment, geography, goal, and many other factors determine the configuration and composition of each insurgent organization. Their composition varies from—
- Organization to organization.
- Mission to mission.
- OE to OE.

Chapter 2

The structure, personnel, equipment, and weapons mix all depend on specific mission requirements. The size, specialty, number, and type of subordinates also depend on the size, number, and specialties required for specific missions in an OE.

Note. Additional details of insurgent direct action and supporting cells are in FM 7-100.4. To find these details, use Army Knowledge Online (AKO) access to the Hybrid Threat Doctrine folder at https://www.us.army.mil/suite/files/30837459. Then, go to the "FM 7-100.4 Org Guide" folder and click, in sequence, as follows: Administration Force Structure; Vol III Paramil Nonmil Orgs; 01 Combatants; 01 Armed Combatants; 01 Insurgent Orgs; 01 Local Insurgent Org. This FM will transition to a TC 7-100.4.

The insurgent organization diagrams of FM 7-100.4 and the personnel and equipment lists that accompany them represent a composite of actual insurgent forces. They are a baseline that U.S. Army trainers can modify to provide the appropriate conditions required for a particular training exercise and/or training task. FM 7-100.4 provides detailed step-by-step instructions on how to construct a task organization based on the training requirements. FM 7-100.4 also describes how to select equipment options. See also TC 7-101 for guidance on creating the appropriate OPFOR order of battle during exercise design.

CAPABILITIES

2-19. Insurgent organizations are flexible, agile, and adaptable. They can quickly change their composition to optimize capabilities and use these capabilities against known or perceived vulnerabilities of an enemy. These dynamic organizations are able to—

- Adjust continually to changing conditions.
- Shift organizational structures and alliances.
- Influence and blend within a relevant population.
- Simultaneously conduct covert and overt actions.
- Use subversion and violence in innovative ways.
- Shift between functional tactics and terrorism.
- Employ a wide spectrum of lethality—from improvised explosive devices (IEDs) to possible WMD.

2-20. The irregular OPFOR can include insurgent organizations in two levels of capabilities:
- Higher insurgent organization.
- Local insurgent organization.

2-21. A higher insurgent organization includes at least one local insurgent organization and can include guerrilla forces. The higher and local insurgent organizations include direct action cells and supporting cells. The direct action cells are primarily within local insurgent organizations. The differences in supporting cells of higher and local insurgent organizations relates to the scope of mission and responsibilities usually associated with the size of geographic areas within which they operate. Some direct action and supporting cells break down into teams.

2-22. A higher insurgent organization can have subordinate guerilla units. In some cases lower-level guerrilla organizations may be subordinate to or affiliated with a local insurgent organization. In either case, guerilla units operate in conjunction with the activities of direct action cells and/or supporting cells.

NETWORKS

2-23. The interactions between and/or among organizations, units, cells, and teams vary in complexity. They can be hierarchical but most often take the form of a network. Communication and coordination in a network can be linear and visualized as links in a *chain*, with each node connected to the next node in sequence. Another option is a *hub*, in which one node (a director or decisionmaker) is central to a number of other nodes that are not directly in contact with each other. A *wheel* is a variation on the hub in which each of the nodes is in contact with neighboring nodes, with the central node providing common source of information and/or guidance. An *all-channel* array provides contact among all nodes of a particular network. The network of an insurgent organization can include a combination of all these types (as shown in figure 2-1). (See figure 2-5 on page 2-14 for another example of such combinations.)

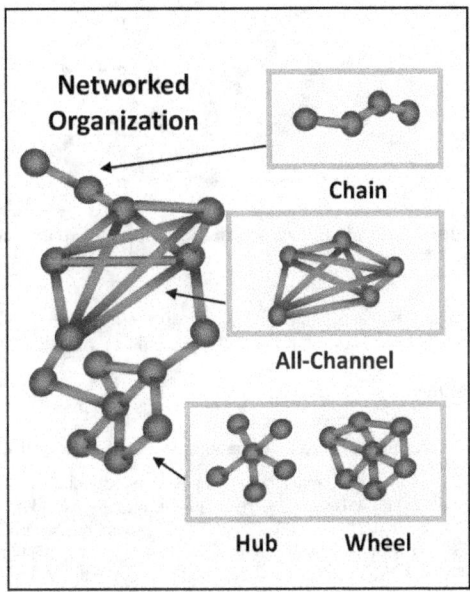

Figure 2-1. Network types

2-24. A network C2 structure adapts configurations to best use the capabilities of available organizations, units, cells, and teams. Adaptability and innovation are hallmarks of the insurgent organization.

LEADERSHIP, ACTION, AND SUPPORT

2-25. Insurgents are flexible in how to organize C2. However, figure 2-2 shows six categories of leadership, action, and support that normally exist in an organizational structure in either a hierarchy or network:
- Senior leaders.
- Subordinate leaders.
- Cadres.
- Active supporters.
- Passive supporters.

2-26. The pyramid illustration in figure 2-2 indicates a relative number of people in each category. Underlying this pyramid is the relevant population that is critical to how these levels function for C2 and/or influence. Insurgents can use the population in various ways, such as a mass human shield or to assist in insurgent organization security.

2-27. The insurgents can also solicit and/or coerce support from the population. Another consideration is the amount of support that is or is not provided by state and non-state sponsors for the insurgency.

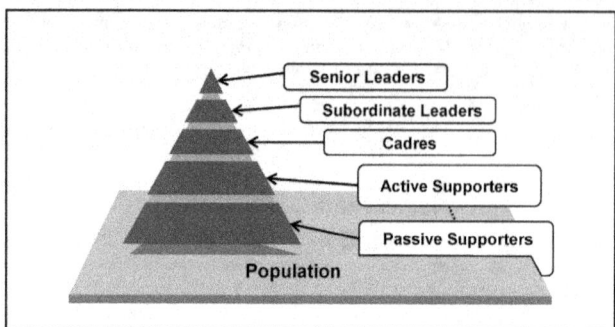

Figure 2-2. Typical levels of leadership, action, and support

Senior Leaders

2-28. Senior leaders are the recognized authority of the insurgency. They conceive and state a philosophical or practical justification for actions by subordinate operational leaders and followers. These senior leaders—
- Announce strategic direction and policy.
- Approve goals and objectives.
- Provide overarching guidance for operations and protracted conflict.

2-29. The leadership of an insurgent organization may be one individual or a group of leaders, who may or may not consult formally with advisors. Leadership may informally use advisors or a council comprised of tribal, religious, political, secular, radical sectarian, other associations, and/or any combination thereof. A formal council of advisors is more common in a higher insurgent organization (see discussion of advisors under that heading, below). The insurgent leader may also be a tribal, cultural, political, or religious leader. The leader of a guerrilla unit may join the leader of a political element in an insurgency and form a collective leadership for the insurgent organization.

2-30. Although leadership may be shared or apportioned among several individuals, one senior leader is often identified as the recognized authority and primary spokesperson for the objectives of the insurgent organization. The leader focuses on the strategic perspective and understands how to communicate the ideology and/or objectives that motivate the insurgents. A leader may have risen from the tactical ranks of the organization with calculated actions and political savvy, and/or may have achieved notoriety and/or demonstrated his influence through acts of terror. The leader usually is knowledgeable of world politics and complex socio-political environments. Individual leadership skills are often supplemented by personal charisma.

2-31. Insurgent leaders can be located within or outside of the insurgent organization's geographic area. These leaders provide strategic guidance that is implemented by subordinate leaders conducting operations within the geographic area. Leaders use the media to convince or coerce targeted segments of a population to act in a particular way. These leaders may solicit clandestine financing and other support for operations to attack high-priority targets.

2-32. A situation may arise where the senior insurgent leadership is coordinating closely with a government-in-exile in order to reestablish authority over a state that has had its legitimate government ousted. Insurgent leaders can use multiple means to support a government-in-exile such as gradually providing civil and social services the existing governing authority is not providing to a relevant population. Another means is to gradually use political processes to obtain legitimate representation in local, regional, and national political parties and institutions. Leaders use these informal and formal processes to gain and sustain support of designated segments of a population and continually expand the positive influence of the insurgent organization's agenda.

Subordinate Leaders

2-33. Subordinate leaders of the insurgent organization command or control organizational activities within geographic areas or functions. These leaders provide direction and guidance, approve goals and objectives, and provide overarching coordination in support of senior leader guidance. They may be leaders of cells and/or teams of specialized expertise or leaders of local insurgent organizations comprising several cells. The subordinate leaders may have responsibilities as special advisors to the senior leadership, media spokespersons, political leaders, or functional experts. If decentralized in structure, these leaders plan and act within general senior leader mission guidance. Other leaders, if structured in a more hierarchical military-type organization, plan and act within specified planning directives or guidance from senior leaders.

2-34. For security and a division of labor, subordinate leaders may sometimes use intermediate leaders to communicate directives and guidance to subordinate cells and supporting activities. Clandestine communications can use sophisticated techniques such as electronic steganography to hide information within a message or email, or simple techniques such as human couriers and intermediaries to relay oral information and/or directives.

2-35. Leaders can have varied effects on an organization's effectiveness if they are eliminated. The insurgent organization develops redundancy among leaders and subordinate leaders in order to minimize significant disruption when a leader is killed or captured.

Cadres

2-36. In a general sense, a cadre is a nucleus of trained personnel capable of assuming control and of training others. In an insurgent organization, cadres include both the political or ideological cadre of an insurgency and the cadre of operators who direct and conduct direct action and supporting functions.

2-37. One type of cadre is the political or ideological core of an insurgent organization. Movements based on religious extremism include religious and spiritual advisors among the cadre. Political advisors may have a special role in some organizations. The cadre can be openly active or mask their activities within a relevant population. Some cadre activities are violent enforcement of political and ideological discipline, subversion of opponents, and/or manipulation of social power to support an insurgent program. Cadre may perform key shadow-government or government-in-exile functions that are physically distant from a relevant population.

2-38. Another type of cadre is the nucleus of experts in direct action or supporting cells of the insurgent organization that carry out specific functions and/or instruct other insurgents in functional capabilities. Some mid-level cadres are trainers and technicians such as bomb makers, financiers, and surveillance experts. Low-level cadres can be direct action operators but usually supervise and coordinate other insurgents rather than actually conducting direct action tasks. These actors can have functional, specialized, multifunctional, and/or supporting purposes based on knowledge and skill sets.

2-39. Examples of activities conducted, coordinated, and led by cadre can include but are not limited to—
- Intelligence and counterintelligence networks.
- Political affiliations and alliances.
- Recruiting programs.
- Training programs.
- Ideology education programs.
- Weapon system and communications acquisition and maintenance.
- Logistics sustainment.
- Finances and fiscal resourcing programs.
- Social and medical welfare support.
- INFOWAR capabilities against the opposition government and adversaries in the relevant population and international community.
- Operational missions.
- Acts of selective and/or random terrorism.

2-40. Some cadres use coercion and leverage to gain limited, periodic, or one-time cooperation from individuals in the relevant population. This forced cooperation can range from gaining specific information on a proposed target to supporting or conducting a suicide bombing attack.

Active Supporters

2-41. Active supporters are fully aware of their relationship to the insurgent organization but do *not* normally commit violent acts. They can continue normal positions in society while providing functional expertise or general support to the insurgents. Depending on the type of organization, they may operate in functional capacities, such as politics, fund-raising, and/or INFOWAR activities of the insurgency. Acting as visible or tacit partners, active supporters may also—

- Conduct or augment intelligence, surveillance, and counterintelligence activities.
- Provide and administer safe houses and safe havens.
- Promote dissatisfaction with the status quo in recurring media affairs information releases.
- Support recruiting efforts for new members and affiliates.
- Solicit and collect financial or other types of donation support.
- Provide civilian transportation.
- Conduct courier and communications services.
- Produce forged documents.
- Acquire commodities to assist insurgent direct action and supporting cells.
- Store caches of supplies, weapons, ammunition, explosives, and other materiel.
- Assist in the manufacture of weapons and IEDs.
- Assist in subversive activities.
- Assist in sabotage, assassination, or other direct and violent actions.
- Provide medical treatment and related support.

2-42. Supporters can improve insurgent initiatives to obtain the representation of legitimately elected political officials. Other supporters may assist the actions of a direct action cell or a guerrilla hunter-killer (HK) team. Active supporters may or may not openly indicate their sympathy or involvement in an insurgent agenda. They can conduct covert and overt operations. They can infiltrate organizations of the governing authority and conduct subversive activities against civil, social, and/or military programs.

2-43. Insurgent leaders often broker associations for active support with civil, political, or other paramilitary organizations for selected purposes. Such associations may arise from specific mutual interests when cooperative efforts might not normally be expected. Public statements and/or secretive agreements that promote mutual support for a specific agenda of an insurgent organization can include the following themes:

- Historical and recent perspectives to remedy grievances.
- Familial, tribal, or clan allegiance.
- Business and social productivity.
- Faith system dogma.
- Personal beliefs, motivations, and ideology.
- Civil and political opposition to programs of a governing authority in a state or region.
- Interference of an extraregional state in the domestic affairs of a state or region.
- Potential future roles in society.

Passive Supporters

2-44. Passive supporters are typically individuals or groups that are sympathetic to the announced goals and intentions of an overarching insurgent agenda. However, they are *not* committed enough to take an active role in insurgent direct actions or acts of terrorism. Sometimes fear of reprisal by opponents of the insurgency leads to passive yet sympathetic support.

2-45. Passive supporters may not be aware of their tacit relationship to the insurgent organization. They may intermingle with active supporters and be unaware of their actual relationship to the organization. Individuals may develop suspicions of activities that might be supporting insurgent actions, but decide to ignore indications and continue their daily lifestyle actions.

2-46. The insurgent organization recognizes that a sympathetic base of passive support in a population is an ever-changing factor with multiple social, economic, religious, or political motives. The insurgency depends on a sympathetic segment of the population remaining passive. Nonetheless, passive supporters can be useful for political activities, fund raising, and/or unwitting assistance in intelligence gathering and other nonviolent activities. A primary value of passive support is minimal interference by the citizenry in ongoing overt actions of the insurgent organization.

2-47. Passive support can undermine civil and social programs, and political and/or theological institutions, of the governing authority. Many functions of such programs and institutions must rely on voluntary cooperation by large segments of the population.

HIGHER INSURGENT ORGANIZATIONS

2-48. The term *higher insurgent organization* includes any insurgent organization at regional, provincial, district, or national level, or at the transnational level. Cities with a large population or covering a large geographic area are considered regions. Higher insurgent organizations may control several local insurgent and/or guerrilla organizations. They can have transnational affiliations or other overt and/or covert support.

2-49. The main capabilities of a higher insurgent organization that are not usually present in a local insurgent organization are as follows:
- Subordinate local insurgent organizations in their network. (Local insurgent organizations usually do not have echeloned subordinate local insurgent organizations.)
- Guerrilla units within their network and operating in the same geographic area. (Local insurgent organizations may have temporary affiliations with guerrilla units or may occasionally command and control them.)
- Personal protection and security cell(s) for the organization's senior leaders and designated advisors or special members. (A local insurgent organization does not normally have a personal protection and security cell.)
- Associations and/or affiliations with criminal organizations directly and without coordination with subordinate local insurgent organizations. (The long-term vision in a higher insurgent organization may include cooperation with regional or transnational criminal organizations for specific capabilities or materiel available through criminal networks.)

2-50. Higher insurgent organizations have no standard organizational structure. They are not necessarily subordinate to a regional, national, or transnational insurgent organization. However, higher and local insurgent organizations can be subordinate to and/or loosely affiliated with regional and national insurgent organizations. Any relationship of independent insurgent organizations to regional or national structures may be one of affiliation or dependent upon only a single shared or similar goal.

2-51. Local insurgent organizations, guerrilla units, and other organizations (such as criminal gangs and networks) often operate in the same geographic area as a higher insurgent organization and can be subordinate, loosely affiliated, or independent of a higher insurgent organization. Any relationship of organizations and units to one another may be allegiance, affiliation, or temporary association based on shared aims and/or mutual support.

2-52. A higher insurgent organization is normally cellular and comprises a network of functional capabilities. Figure 2-3 shows the types of organizations and cells that can comprise a higher insurgent organization.

Chapter 2

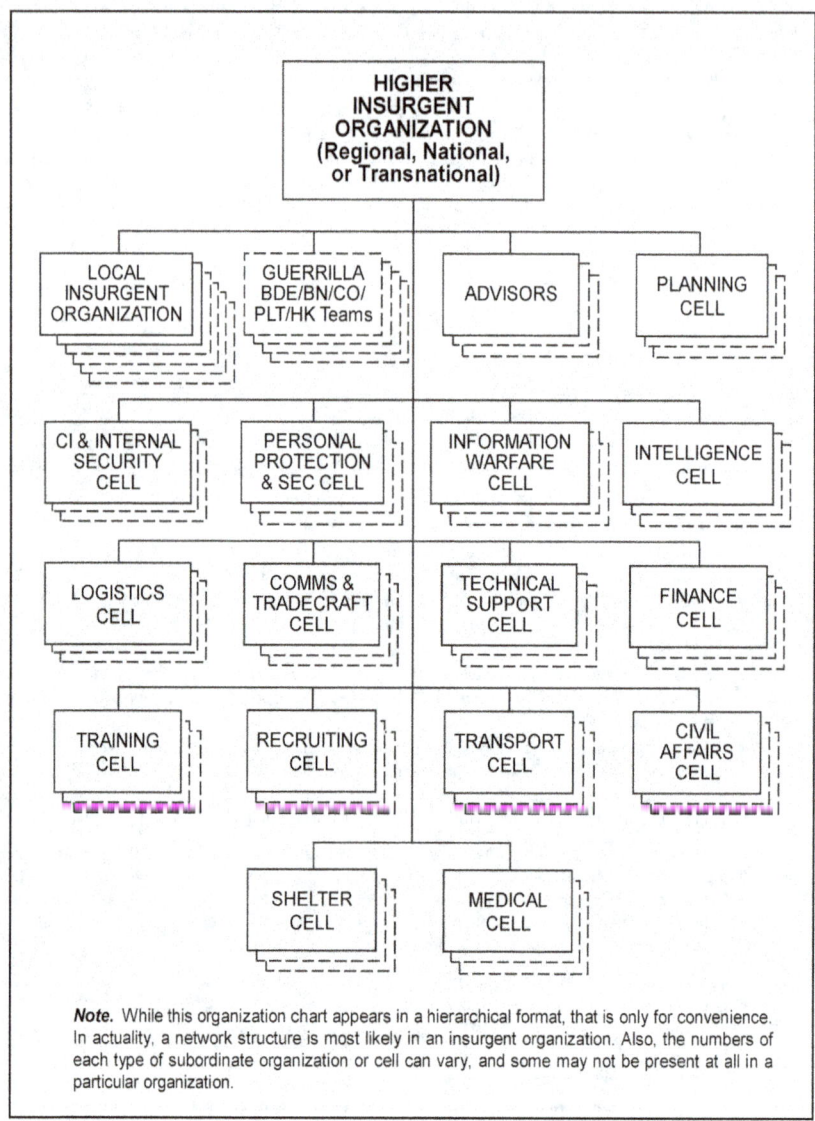

Figure 2-3. Higher insurgent organization

2-53. Some insurgent organizations may not have separate cells for all of the required functions. However, someone in the organization must be designated to perform all basic and critical functions. For example, the intelligence cell may also incorporate the counterintelligence and security functions, or the logistics cell may also perform the functions of the shelter and/or transportation cells.

2-54. Subordinates and supporting cells of a higher insurgent organization may or may not be colocated with each other or with the leadership of the higher insurgent organization. They may be located in separate villages, towns, cities, regions, or countries, as long as they can adequately and securely provide the required support in a timely fashion. The numbers and types of each, and their locations, may change frequently.

2-55. A higher insurgent organization can have all the same types of supporting cells found in a local insurgent organization. (These types are described under Local Insurgent Organization.) Compared to their local counterparts, these cells may be larger and have expanded capabilities. However, the higher organization may also have advisors and one or more personal protection and security cells.

Advisors

2-56. The senior leader (or leaders) may or may not have a council of advisors. He can receive advice from others inside and outside the organization. However, he may rely heavily on his council of advisors to provide counsel, guidance, and/or expertise on a wide variety of areas. If present, advisors can include—

- Political (international relations; shadow government).
- Military (advice and liaison with affiliated nations or guerrillas; acquisition of military weapons and equipment, advisors, or fighters).
- Religious (religious leaders or liaison with them).
- Cultural (ethnic, tribal, minority populations).
- Financial (international and money-raising).
- Mass media (perception management).

2-57. The number and types of advisors is situation dependent. Some insurgent leaders may have an advisor for every issue requiring specialized expertise, while other leaders may have advisors that can provide counsel on several functions. An example of a multifunctional advisor would be one that could provide counsel on political, religious, and tribal issues. Some functions may require specialized expertise. For example, in order to manipulate the media to achieve a specific goal in perception management, the leader of the INFOWAR cell may require expertise and/or advice from several advisors such as a cleric, a politician, and a tribal and/or ethic leader.

2-58. The number of permanent advisors can be many but is often limited to a few individuals. One of these advisors generally rises to the position of primary advisor to the leader. In some insurgent organizations, in addition to providing counsel, the primary advisor also serves in a position comparable to that of a second-in-command, a deputy, or a chief of staff. The position of primary advisor is an extremely prestigious, influential, and powerful position in the insurgency. The primary advisor is usually the religious, political, or the military advisor. The choice of primary advisor typically reflects the insurgency's goals and agenda.

2-59. Advisors may consult with the leader on policies, tactics, and weapons to be used. They may use simple or information-age technology to exchange experience, practical observations, and lessons learned. Such advisors may coordinate activities among various supporting cells while also coordinating selective activities of direct action cells.

2-60. Even if not part of a formal council of advisors, leaders of supporting cells directly assist a senior leader in planning and supervising the actions of the insurgent organization. For instance, the leaders of the INFOWAR, logistics, and finance cells may advise the leader on matters pertaining to their functional specialties.

2-61. Advisors may or may not be armed. In most cases, advisors will possess firearms but may not choose to carry them on all occasions. There are typically security specialists (personal bodyguards) and drivers assigned to the advisors. The advisors' association with the insurgency may be covert, with very few people aware of their connection. Most insurgent organizations have a mixture of overt and covert counselors.

Personal Protection and Security Cells

2-62. A personal protection and security cell is responsible for the personal security and welfare of the senior leader(s) and other important persons. The cell's armed security specialists provide around-the-clock protection for these personnel and vehicles for transporting them. The vehicle may or may not be armored. Even if armored, it is built to appear just like any other vehicle commonly used in the environment. Some weapons and/or equipment may be left in the vehicle until required or not carried on the mission at all. All cell members are cross-trained in the use of all weapons, equipment, and vehicles assigned to the cell.

2-63. This cell works closely with both the counterintelligence and security cell(s) and the intelligence cell(s). When needed, it receives augmentation of trained personnel from direct action teams.

LOCAL INSURGENT ORGANIZATIONS

2-64. The term *local insurgent organization* applies to any insurgent organization below regional, provincial, or district level. This includes small cities, towns, villages, parishes, communities, neighborhoods, and/or rural environments. (Large cities are equivalent to regions and may contain several local insurgent organizations.) Activities remain focused on a local relevant population.

2-65. Differences between a local insurgent organization and a higher insurgent organization are as follows:
- *Direct actions cells* are present within a local insurgent organization. Their multifunctional and/or specific functional capabilities may be enhanced or limited based on availability of resources and technical expertise in or transiting the local OE. These direct actions are planned for immediate and/or near-term effects related to the local insurgent organization's area of influence.
- *Guerrilla units* might not be subordinate to the local insurgent organization. However, temporary affiliations between local insurgents and guerrillas are possible for specified missions coordinated by a higher insurgent organization. Direct action personnel may use, fight alongside of, or assist affiliated forces, and guerrillas to achieve their common goals or for any other agenda. Guerrilla units may operate in a local insurgent organization's area of influence and have no connection to the local insurgent organization or a higher insurgent organization.

2-66. *Criminals* can affiliate with a local insurgent organization or a higher insurgent organization as a matter of convenience and remain cooperative only as long as criminal organization aims are being achieved. The local insurgent organization retains a long-term vision of its political agenda, whereas cooperation by a criminal organization is usually related to localized commercial profit and/or organizational influence in a local environment. This usually equates to criminals controlling or facilitating materiel and commodity exchanges. The criminal is not motivated by a political agenda.

2-67. The local insurgent organization uses functional tactics (see chapter 7) and terrorism (see chapter 6) as the primary means to achieve its goals. Terrorism instills fear and anxiety that coerces and degrades the resolve of an enemy governing authority and selected people in a relevant population.

Relation to Other Insurgent Organizations

2-68. The local insurgent organization is the basic level of insurgent organization. Local insurgent organizations are not always subordinate to a regional, national, or transnational insurgent organization. They may be completely autonomous and independent of a larger insurgent movement and not be associated with it in any way. In other cases, they can be either subordinate to or loosely affiliated with such a larger organization. They may operate under the guidance of a larger insurgent organization even is no command relationship exists. In some cases a local insurgent organization may provide only financial support and general guidance to its direct action and supporting cells (see figure 2-4).

2-69. Cells of a local insurgent organization may be forced to provide for themselves in several areas. A typical example of this is a smaller direct action cell separated from the parent insurgent organization by distance, population, or ability to communicate securely. They may not have access to the expertise or products such as IEDs provided by the technical support cell and must improvise IEDs by themselves.

2-70. Any relationship to a higher organization or among independent local insurgent organizations may be dependent upon only a single shared or similar goal. These relationships are generally fluctuating and may be fleeting, mission dependent, event- or agenda-oriented, mutually coordinated, and/or coerced for a specific temporary purpose. There may be loose coordination of certain actions, after which the organizations revert back to their independent modes.

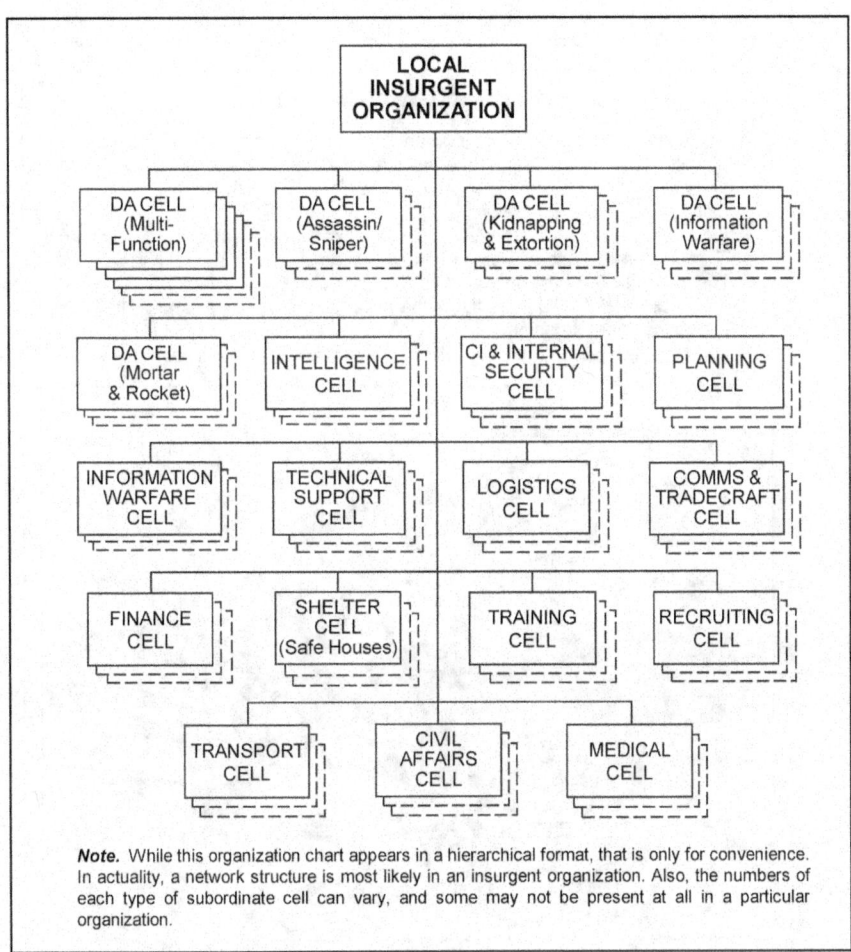

Figure 2-4. Local insurgent organization

Dispersion

2-71. Direct action cells and supporting cells disperse as a norm. They are not normally colocated with each other or with the leader(s) of the local insurgent organization. Although in some cases direct action cells may come together to provide mutual support, they usually carry out their mission independent of other cells. Dispersion enhances the security and survivability of the local insurgent organization and its individual cells. It also distributes the influence of the insurgency to a wider area and a larger segment of the population.

Leadership

2-72. Leadership in a local insurgent organization may reside in one individual or more than one individual. Advisors or a small council may exist but are not always present in a local insurgent organization to assist the leader or leaders. (See Advisors under Higher Insurgent Organization, above.) In some cases, command directives can be very specific and focused. In other cases, the leadership may purposely issue only general guidance to allow initiative at the cellular level for specific actions.

Insurgent Network Example

The C2 structure of a local insurgent organization can be a network, hierarchy, and/or a combination of networked and hierarchical systems. However, a network is the most likely structure. An example of a network in figure 2-5 includes—

- A higher insurgent organization.
- A local insurgent organization (subordinate to the higher insurgent organization).
  - Direct action cells (some broken down into teams).
  - Supporting cells (some broken down into teams).
- Guerrilla units (subordinate to the higher insurgent organization and affiliated with the local insurgent organization).
- A criminal organization (affiliated with the local insurgent organization).

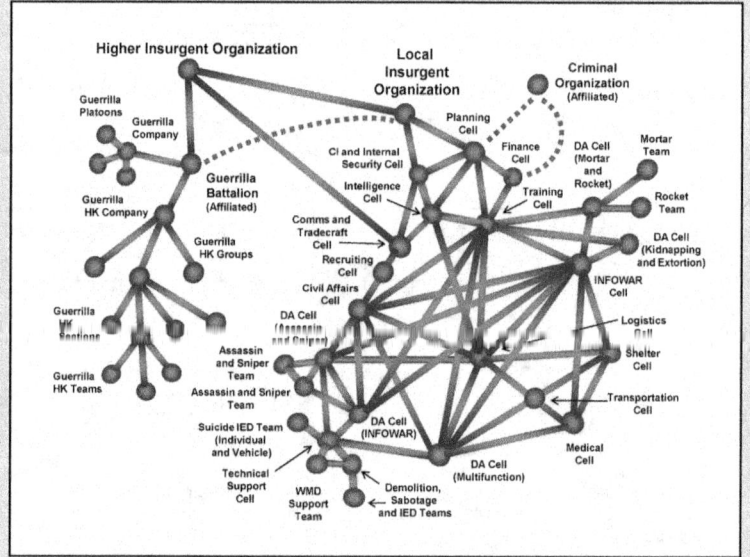

In this example, the local insurgent organization appears to be self-sufficient, relying on its own supporting cells for support. In some cases, supporting cells of the local insurgent organization would also have links to their counterpart cells in the higher insurgent organization. In other cases, the local insurgent organization may lack some supporting cells if it relies on the higher organization for this support.

Figure 2-5. Local and higher insurgent organizations network (example)

Cellular Structure

2-73. As with a higher insurgent organization, local insurgent organizations have a cellular structure. However, there are several differences in functions and scope of operations. Figure 2-4 and 2-5 show the types of cells that may be present in a local insurgent organization. Local insurgent organizations are typically composed of anywhere from 3 to more than 30 direct action and supporting cells. However, a smaller insurgent organization may consist of as few as one direct action cell (see figure 2-6).

2-74. Cellular structure in the local insurgent organization comprises two basic types of capability. These cells are—
- Direct action cells.
- Supporting cells.

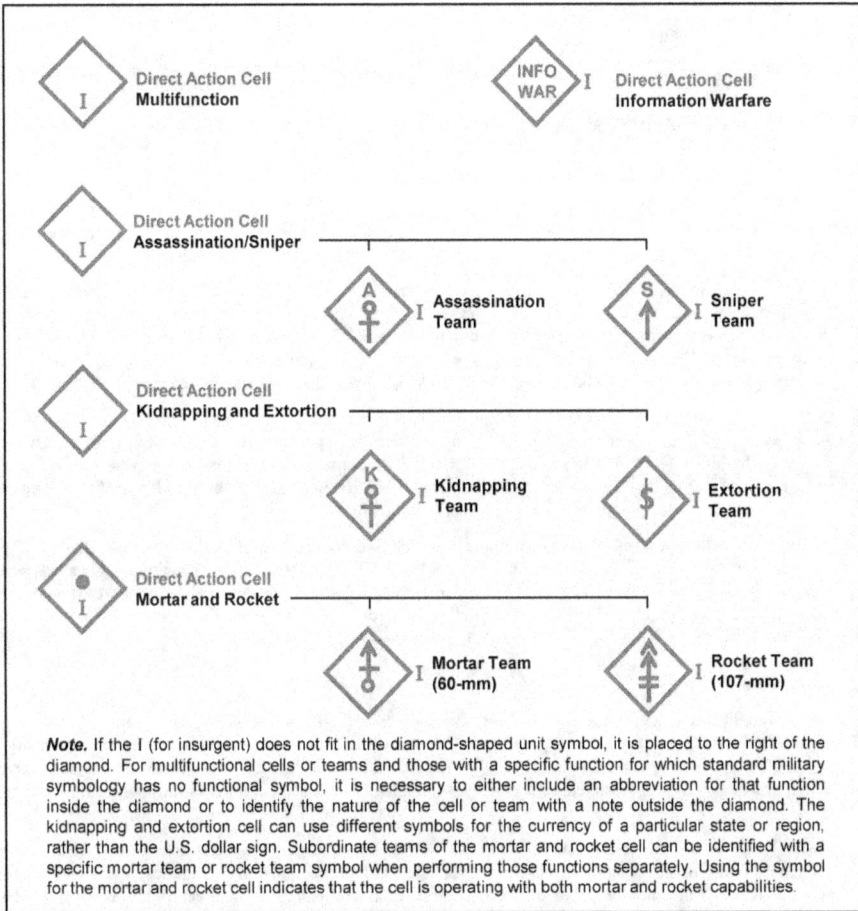

Figure 2-6. Insurgent direct action cell and team graphic symbols

Direct Action Cells

2-75. *Direct action* cells are the backbone of the insurgency movement. They contain the actual fighters of the insurgency. Some direct action cells have a *multifunction* purpose, while others are organized to perform a *specialty* function. Specialties can include, but are not limited to, assassination, sniper, kidnapping, extortion, INFOWAR, or mortar and/or rocket attacks. Every direct action cell is capable of performing all the missions listed above, to some degree. All cells are capable of sabotage and IED emplacement and detonation.

2-76. Typical types of direct action cells that may be present in a local insurgent organization are—
- Multifunction.
- INFOWAR.
- Assassination and sniper.
- Kidnapping and extortion.
- Mortar and rocket.

2-77. The types of direct action cells and numbers of each can vary greatly from one local insurgent organization to another. Some may have—
- All multipurpose cells.
- All specialized cells, each with a different focus.
- All specialized cells with a single focus.
- Any mixture of the above.

Most insurgent organizations have a mix of separate specialty cells and multifunction cells. All these cells are designed to operate independently (if necessary) once they have started their mission.

2-78. Direct action cells do not have a fixed structure. The structure, personnel, equipment, and weapons mix all depend on specific mission requirements. The size, specialty, and number of cells employed also depend on the size, number, and specialties required for specific missions. A cell typically contains from 6 to 10 personnel. However, a cell can be as small as 2 people or consist of over 20 people. Direct action cells easily lend themselves to be broken down into subordinate teams whenever necessary.

2-79. All direct action cells record key events and successful results on digital video and/or still cameras. In some cases, the key event may be staged for the camera. Upon mission completion, the direct action cells turn the information over to the INFOWAR cell for manipulation and exploitation. The videographers or camera operators try to blend in with crowds and disassociate themselves from the operation. Sympathizers in the local population may also serve in this role.

2-80. Direct action personnel may be a mixture of men, women, and children. Local women and children may be used as runners, messengers, scouts, guides, drivers, porters, snipers, lookouts, videographers, camera operators, or in other roles. They may also emplace and/or detonate IEDs, signal flares, and mines. Women (and possibly children) may be fighters and participate in assassinations, ambushes, or assaults. They may also serve (willingly or unwillingly) as suicide bombers.

Multifunction Cell

2-81. The direct action cell (multifunction) is the basic direct action cell structure. It can employ functional tactics or terrorism, as necessary. It possesses all the capabilities of specialized cells (such as assassination and sniper; kidnapping and extortion; INFOWAR; and mortar and rocket) but normally to a lesser degree. When not engaged in specialized activities the, specialized cells can serve as multifunction direct action cells.

2-82. Bombs (especially IEDs) are the weapon of choice for multipurpose direct action cells. They can be used in support of assassination, maiming, sabotage, and producing mass casualties. The cells usually acquire IEDs from the technical support cell as unassembled, pre-manufactured components. The multifunction cell assembles them and adds fuzes and detonators. Within the multifunction cell, an IED team usually includes at least three people: a lookout, the IED emplacer, and a triggerman. In some cases, a small multifunction cell of three or four personnel may act as an IED cell. The team or cell emplaces the

IEDs, and the triggerman detonates them at the appropriate time. If additional assistance or IED expertise is required, they receive it from the technical support cell.

2-83. Insurgents often use IEDs as secondary devices, to detonate on the arrival of personnel responding to another attack or IED. IEDs can be detonated by a variety of means, including remote, command, electrical, trip wire, pressure, time, and others.

Assassination and Sniper Cell

2-84. The primary mission of the direct action cell (assassination and sniper) is to terrorize a relevant population and/or to assassinate preselected persons. (See chapter 6 on Terrorism.) In either the assassination or the sniper task, the cell creates a psychological impact to intimidate and demoralize the population. The cell may indiscriminately select individual targets in crowded marketplaces or religious and political gatherings with the sole intent to terrorize.

2-85. An assassination is a deliberate action to kill political leaders or very important people rather than killing common people, which is considered murder. The insurgent assassinates or murders people it cannot intimidate, who have left the group, or who have some symbolic significance for the enemy or world community. Insurgent organizations may refer to these killings as "punishment."

2-86. Many targets of assassination are symbolic and often have a great psychological impact on the enemy. For example, assassinating an enemy negotiator or successful businessperson can demonstrate the enemy's inability to protect its own people.

2-87. The assassination and sniper cell consists of two teams. The cell leader serves as a team leader of one team, and another insurgent serves as the assistant cell leader and team leader of the other team. The structure, personnel, equipment, and weapons mix all depend on specific mission requirements. Weapons appropriate to each mission are selected. The other equipment is left behind or cached.

2-88. The cell could select its own targets based on opportunity, as long as hitting such targets supports (or does not conflict with) the overall goals of the insurgent organization. Assassination methods include remotely-detonated bombing, the use of firearms, and poisoning. The target's vulnerabilities determine the method of assassination. The insurgent leadership and/or the planning cell can exercise centralized control as necessary in order to orchestrate concerted action or to achieve a specific goal in its area of influence.

2-89. When not engaged in specialized activities, the assassination and sniper cell serves as a multifunction direct action cell. Depending on mission requirements, the assassination and sniper cell may also serve as a reconnaissance element, a stay-behind element, or provide security for other direct action cells.

Kidnapping and Extortion Cell

2-90. The primary mission of the direct action cell (kidnapping and extortion) is to conduct kidnapping, extortion, hijacking, and hostage-taking. These acts help to finance the insurgent organization and serve to influence and/or terrorize the population. (See chapter 6 on Terrorism.) Each action, particularly a kidnapping, requires detailed planning and support by a number of other cells.

2-91. The kidnapping and extortion cell may be authorized to identify victims as targets of opportunity as long as hitting such targets supports (or does not conflict with) the overall goals of the insurgent organization. However, the insurgent leadership can exercise centralized control as necessary, in order to orchestrate concerted action, and might identify certain targets or types of targets in order to achieve a specific goal in its area of influence. For a mission directed by the leadership, this cell may receive augmentation of insurgents from other cells.

2-92. The kidnapping and extortion cell leader coordinates with the intelligence cell for reconnaissance and surveillance of targets. The intelligence cell provides information on the target's vulnerabilities, weaknesses, and routine behavior. Every member of the insurgent organization as well as sympathizers can help gather this target information. The intelligence cell must thoroughly analyze the target so that it can advise the insurgent leadership, the planning cell, and/or the kidnapping and extortion cell on selection of the target site and method of kidnapping. Human intelligence is vital when planning and conducting a kidnapping. The intelligence cell continuously conducts surveillance on the target to identify potential security flaws and identifies vulnerabilities that the direct action cell can exploit.

2-93. The intelligence cell and/or the kidnapping and extortion cell conducts detailed reconnaissance/surveillance of potential kidnap sites to determine the best site, considering cover, concealment, and escape routes. Once the target and site are selected, the kidnapping and extortion cell (and/or the planning cell and intelligence cell) conducts detailed analysis to provide the kidnapping and extortion cell with the requisite data on the target. The information required depends on the location of the target and site, but typically includes—

- The exact route the target uses.
- The method of conveyance and its specific characteristics.
- The number of security personnel, their location, disposition, and types of weapons used.
- The target's likes, dislikes, allergies, habits, and routines.

2-94. After planning and reconnaissance, the kidnapping and extortion cell conducts the actual action. For a kidnapping, the cell leader designates a specific team to conduct the kidnapping. After receiving intelligence, the team rehearses specific kidnapping techniques, such as an ambush or abduction, and finalizes planning. The team plans the escape route in great detail because of the complexities of transporting the victim. It usually disables the victim to make the escape easier. The team determines the best method of disabling the victim (such as drugging, stunning, or binding him).

2-95. The kidnapping and extortion cell coordinates with the shelter cell for a safe house when kidnapping and/or hostage-taking is the mission. The INFOWAR cell helps create and maintain the fear caused by kidnapping and extortion through its propaganda and media manipulation means.

Information Warfare Cell

2-96. The direct action cell (INFOWAR) supports the insurgent organization's INFOWAR plan. It may or may not receive guidance and assistance from the local insurgent organization. INFOWAR cell or its counterpart in a higher insurgent organization. (See appendix A for more detail on INFOWAR.)

2-97. There may be as few as one direct action cell (INFOWAR) to over 20 such cells depending on the mission, the targets within a relevant population and governing authority, and topics to be exploited. The cells can be broken down into teams when necessary. The structure, personnel, equipment, and weapons mix, all depend on specific mission requirements. Personnel select weapons appropriate to the mission. Other weapons and equipment are added as required, such as computers, computer-rigged vehicles, specialized antennas, and communications.

2-98. Depending on the size, nature, and focus of the insurgent organization, the cell may be capable of several functions. Examples of functions performed by the INFOWAR cell are—

- Selective sabotage actions.
- Information management (internal methods, links, and security).
- Media manipulation (misinformation and disinformation).
- Communications (cyber embeds via Internet sites, propaganda and indoctrination videos; broadcast successes of the direct action teams).
- Civic actions popular to a relevant population.
- Cyber-mining for information and intelligence.

Some of the functions may require specialized expertise. For example, the media manipulation function may require expertise and/or advice from a cleric; a political, tribal, ethic, or cultural leader; or other subject matter experts. Some functions can also be performed by personnel outside of the INFOWAR cell. Several members of the cell may be hired INFOWAR specialists or "gun fighters."

Mortar and Rocket Cell

2-99. The primary purpose of the direct action cell (mortar and rocket) is to either terrorize or influence the local populace and governing authorities. It may indiscriminately fire into crowded marketplaces and religious gatherings with the intent to terrorize and influence a relevant population. Targets may include religious or national icons and/or landmarks. These attacks are often used to support the overall INFOWAR plan. They may also be used in a manner to shift blame for the attack to the enemy. The cell can provide indirect fires in support of insurgent missions. When insurgents need additional mortar and rocket fires, they may look to affiliated guerrillas for support.

2-100. The rocket and mortar cell typically consists of a mortar team and a rocket team. Each team may have approximately six personnel. The mortar team of the cell may have light and/or medium mortars. When it uses medium mortars, it may require additional ammunition bearers and possibly a light truck for transporting the weapon. The rocket team normally fires medium rockets, sometimes from improvised rocket launchers. The mortar team and rocket team can conduct missions as a cell or conduct mortar and rocket missions as separate teams. The normal plan is for the team(s) to execute a fire mission and quickly disperse from the firing point.

2-101. When not engaged in specialized activities, the mortar and rocket direct action cell can serve as a multifunction direct action cell. The mortars, rockets, and associated equipment may be cached or left behind, in which case the cell members carry weapons and munitions similar to the multifunction cell.

Supporting Cells

2-102. *Supporting cells* support and assist operations of the direct action cells or the insurgent organization as a whole. At the local level, the supporting cells either support the roles of the direct action cells or exploit their successes. The same types of supporting cells may be present in a higher insurgent organization. Figure 2-7 shows the graphic symbols of supporting cells.

2-103. Supporting cells that may be present in the insurgent organization are—
- Intelligence.
- Counterintelligence and internal security.
- Planning.
- INFOWAR.
- Technical support.
- Logistics.
- Communications and tradecraft.
- Finance.
- Shelter.
- Training.
- Recruiting.
- Transportation.
- Civil affairs.
- Medical.

2-104. The local insurgent organization normally possesses the supporting capabilities listed here. However, some of these functions may be combined, rather than having separate cells. Also, supporting cells are not limited to these types.

Intelligence Cell

2-105. The intelligence cell plans, coordinates, and implements the insurgent intelligence collection plan and provides intelligence and information to support insurgent operations. The cell also conducts reconnaissance to obtain information about the activities, tactics, and resources of the enemy and potential supporters of the insurgency. Reconnaissance methods include surveillance, use of informants, and infiltration of organizations. Observation is the most common method used to conduct reconnaissance.

2-106. The insurgent organization usually produces its own general intelligence and targeting information. Every member of the insurgent organization is an intelligence-gathering mechanism. Intelligence cell personnel may serve in any occupation (such as a taxi or delivery driver, or truck driver) that allows them to blend in with the population and still provides them the flexibility and mobility needed

Chapter 2

to gather information. Information in raw form may be freely provided by sympathizers conducting surveillance on behalf of the insurgent organization while living, traveling, or working near either a target area or the enemy. Raw information may also be purchased locally from affiliated insurgents, guerrillas, or criminal organizations. The intelligence cell analyzes all this information and turns it into intelligence.

2-107. In addition to conducting intelligence analysis, typical activities of an intelligence cell may include—

- Tracking enemy movements.
- Determining enemy tactics, techniques, and procedures.
- Scouting potential targets.
- Establishing enemy vulnerabilities.
- Selecting attack locations.
- Stalking potential assassination targets.

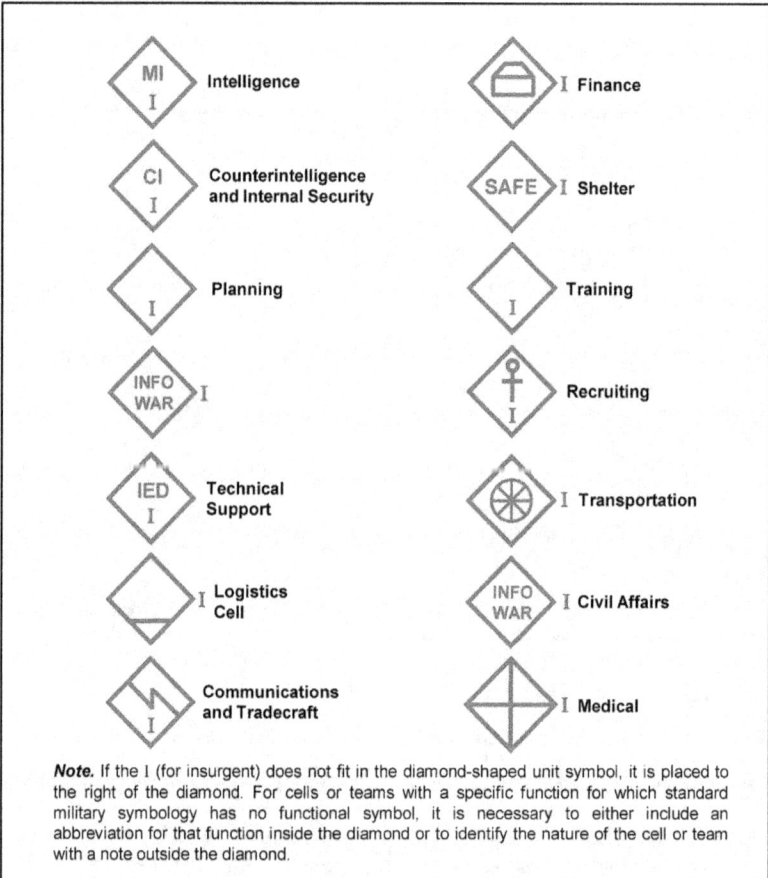

Figure 2-7. Insurgent supporting cell graphic symbols

2-108. Collecting information is a continuous function performed by every insurgent organization. Overt activities include the open collection of information by individuals who circulate among the people. Other activities involve secret collection of information. This can include information collected through the use of extortion, bribery, or coercion. Insurgents also collect information using electronic devices and human-intelligence agents who may join or infiltrate popular organizations, government organizations, and nongovernmental organizations. The intelligence cell of a local insurgent organization may further disseminate the information to its counterpart intelligence cell in a higher insurgent organization.

2-109. When planning any action, an intelligence cell analyzes information from all available sources. Sleeper agents, members of the insurgent organization, or active supporters and sympathizers who may have resided within the target area for years often have the specific mission of gathering information. This information may later serve to support direct action missions.

2-110. The information gathered by the insurgent organization is used to plan future activities and determine the feasibility of planned actions. In the offense, efforts are concentrated on the enemy at his location and the area surrounding the target. In the defense, the reconnaissance effort is to determine when and where the enemy will conduct offensive actions against insurgent forces.

2-111. The insurgent organization's intelligence may be superior to the enemy's. This is because insurgents may have—
- More intimate first-hand knowledge of all aspects of the local environment.
- Close relationships with the population.
- Penetration of governmental structures.
- Ability to maintain observation across the countryside or urban area.

Counterintelligence and Internal Security Cell

2-112. The counterintelligence (CI) and internal security cell conducts internal CI and operations security (OPSEC) activities. Members of this cell infiltrate other cells to identify security weaknesses or breaches, and enforce proper security measures. The cellular structure of the insurgent organization helps ensure against the compromise of the identity, location, or activities of leaders and members of other cells should there be a breach of internal security.

2-113. The local insurgent organization must place great emphasis on the conduct of internal CI activities because of the criticality of maintaining cohesion within the hazardous environment in which it operates. It is susceptible to infiltration by enemy agents. If the local insurgent organization is infiltrated, it will not survive.

2-114. It is within the CI and internal security cell that all OPSEC measures for the organization's activities, as well as other security measures, are developed, disseminated, and enforced. In most insurgent organizations, violation of these security rules can result in immediate death to the violator and/or his family members. This cell is responsible for maiming or assassinating current or former insurgent members who commit breaches. Paranoia among the members actually increases security, since all members desire to remain free of suspicion. Individuals assigned to the CI and internal security cell are usually mature, experienced, or senior in the organization.

Planning Cell

2-115. The planning cell conducts near-, mid-, and long-term planning for missions of the insurgent organization. This cell plans current and future actions of direct action and other supporting cells that contribute to the goals of the overall organization. It coordinates cooperation among cells, when necessary. It works closely with the intelligence and training cells to coordinate capabilities and minimize the impact of organizational limitations on insurgent operations.

2-116. The cellular structure of the insurgent organization helps ensure against the compromise of the plans and actions of direct action and/or supporting cells should there be a breach of internal security. Communication between and among cells may use intermediaries to provide additional security from infiltration and/or identification of key leaders and planners.

Chapter 2

Information Warfare Cell

2-117. The information warfare (INFOWAR) cell plans, coordinates, and implements the local insurgent organization's INFOWAR activities. (See appendix A for more detail on INFOWAR.) This cell also provides guidance and assistance to the direct action cells (INFOWAR) whenever required. Portions of the cell may be dispersed.

2-118. Depending on the size, nature, and focus of the insurgent organization, the INFOWAR cell may be capable of several functions. Some example functions are—

- Information management (internal methods, links, and security).
- Media manipulation (misinformation and disinformation).
- Public affairs designed to influence the population
- Communications (cyber embeds via Internet sites, propaganda videos, broadcast successes of direct action teams, and printing).
- Rumor control.
- Sabotage actions.
- Civic actions.
- Fund-raising (including international).
- Recruiting.
- Indoctrination training.
- Assistance in cyber-mining for intelligence.

All these functions are integrated to further short- and long-range goals. Some of the functions may require specialized expertise. For example, the media manipulation function may require expertise and/or advice from a cleric; a political, tribal, ethic, or cultural leader; or other subject matter experts.

2-119. IED and other attacks are often used to support the overall INFOWAR plan. In all direct action cells, key events (especially successes) are recorded on digital video and still cameras. Upon mission completion, the recordings are turned over to the INFOWAR cell for manipulation and exploitation. Although the direct action cell has a video camcorder and digital camera capability within its cell, an INFOWAR cell may be tasked to provide video or digital collection of the event for rapid media distribution in the insurgent INFOWAR campaign.

Technical Support Cell

2-120. The technical support cell is responsible for all acquisition, manufacturing, and storage of bombs, mines, and other tactical demolitions and fuze materiel. It is the primary bomb factory for the insurgent organization. Bombs can include—

- IEDs.
- Signal flares (actually a subcategory of IEDs).
- Bombs configured for individual and vehicular suicide tasks.
- Even a WMD, when materiel and expertise is available in the insurgent organization.

2-121. Bombs (specifically IEDs) are a preferred weapon of choice for the insurgent organization. This is because—

- They are inexpensive.
- The materials are readily available.
- They are relatively easy to build.
- They are extremely destructive.

Bombs (IEDs) may be very sophisticated or extremely simple. They easily lend themselves to terrorizing the population. They are used in support of assassination, maiming, sabotage, mass casualties, and mass destruction.

2-122. Although members of the technical support cell are very capable of emplacing and detonating IEDs and other demolitions, they normally do not do so. Their expertise is far too critical for them to routinely go on direct action missions. Direct action cells emplace and trigger the devices. On occasion, a representative from the technical support cell may be required to accompany the direct action cell to properly emplace and detonate the device, especially when dealing with WMD IEDs. Those insurgents requiring specific expertise in the fabrication, emplacement, and detonation of WMD such as radiation (dirty bombs) and biological weapons may acquire the expertise and material from outside the local insurgent organization.

2-123. The number of teams subordinate to the technical support cell is not fixed and varies depending on the specific missions. Types of teams that may comprise the technical support cell are—
- Demolition, sabotage, and IED team.
- Suicide IED team (individual and vehicular).
- WMD support team.

2-124. **Demolition, Sabotage, and IED Team.** The demolition, sabotage, and IED team is the primary IED and tactical munitions factory for the technical support cell. This team is responsible for all acquisition, manufacturing, and storage of IEDs, suicide bombs, side-attack (antitank and anti-vehicle) mines, and other tactical demolitions and fuzing for munitions. The team prepares these devices for distribution to other elements of the insurgent organization or affiliated organizations and/or persons. Some demolition, sabotage, and IED teams may be located in factories in small villages (or other remote areas or local accommodations) where they build their IEDs and then smuggle them into cities, where suppliers may then distribute them (or sell them) to insurgent organizations or other customers.

2-125. This team provides IEDs (usually unassembled, pre-manufactured components) to direct action cells. The direct action cell then assembles and adds a fuze or detonator to the IEDs. Some direct action cells may not have access to the expertise or products (IEDs) provided by the technical support cell. In these cases, the team trains, advises, and provides expertise to direct action cells. The team may also train direct action cells to manufacture limited qualities of IEDs for their own use.

2-126. The demolition, sabotage, and IED team may train, advise, and provide expertise to direct action cells (especially the multifunction cells) on how, where, and when, to emplace and detonate munitions and on the proper assembly, fuzing, and detonation of the devices. The team also provides instruction on remotely detonated mines.

2-127. **Suicide IED Team.** A suicide IED team (individual and vehicular) is responsible for acquisition, manufacturing, and storage of IEDs and/or vehicle borne improvised explosive devices (VBIEDs) for suicide missions. The number of such teams in a technical support cell is not fixed and varies depending on specific missions. All suicide IED/VBIED team personnel are cross-trained to perform all functions necessary to fabricate IEDs and VBIEDs including automotive welding and fabrication. Members of the team are very capable of detonating IED/VBIEDs and other demolitions, but they normally do not do so—even to detonate a device carried or driven by someone else.

2-128. Direct action cells are responsible for the emplacement and detonation of the IED/VBIEDs. Suicide bombers/drivers are recruited by the recruiting cell, and turned over to direct action cells to properly emplace the individual-carried IEDs and VBIEDs. Once the direct action cell has ensured the proper emplacement of the IED/VBIED, they can either trigger the IED/VBIEDs remotely or use other detonation methods. On occasion, a representative from the suicide IED/VBIED team may be required to accompany the direct action cell to ensure proper emplacement and detonation of the device.

2-129. **WMD Support Team.** The WMD support team has the primary purpose of creating weapons to either terrorize or influence a relevant population and governing authorities. Insurgent organizations use indiscriminate techniques to create mass casualties and mass disruption. Chemical, biological, radiological, and nuclear (CBRN) weapons are the potential weapons of choice for organizations employing terror tactics. Insurgent organizations may obtain or use CBRN weapons for a variety of motives. They might threaten the use of these weapons as "saber rattlers" to raise the ante in response to political or military actions of the governing authority or foreign powers. They may actually use such weapons to achieve a specific objective or to terrorize.

2-130. Use of a WMD such as a radiological (dirty bomb) and biological weapon can require specific expertise in fabrication, emplacement, and detonation or activation of the WMD. The WMD support team acquires necessary expertise and material from outside the local insurgent organization. Other types of WMD may be relatively simple to fabricate and use with devastating effects.

2-131. Toxic industrial chemicals (TICs) and/or toxic industrial material (TIM) can be used as a WMD to cause mass casualties and psychological stress and anxiety. Most WMD support teams have easy access to TICs and TIM. TICs are highly toxic commercial chemical substances with acute toxicity that are produced in large quantities for industrial purposes. They can be solid, liquid, or gas. These are the normal weapons of choice for the WMD support team. TIM has similar toxicity and can be used as a complement to TICs or other mass effects on an enemy such as officials of a governing authority and/or a relevant population. Even though a mass release of TICs or TIM causing numerous noncombatant casualties was due to an IED explosion, the insurgent organization might be able to blame the casualties on the enemy as an intentional act.

2-132. Once an insurgent organization acquires the ability to produce WMD IEDs, a WMD support team will normally have at least one such weapon fabricated and available at any given time. It may also have others in various stages of manufacture.

2-133. All WMD support team personnel are cross-trained to perform all functions necessary to fabricate WMD IEDs. However, team members do not normally emplace and detonate WMD IEDs because their expertise is too critical to risk in direct action missions. Direct action cells emplace and trigger the WMD IEDs. On occasion, a representative from the WMD support team may need to accompany the direct action cell to ensure proper placement and detonation of the weapon.

Logistics Cell

2-134. The logistics cell is the resource planner for the insurgent organization. This cell is responsible for the planning, acquisition, and distribution of all resources—human and materiel. Resources can come from sources outside of the insurgent area of interest, or can be obtained from within the geographic area. In addition to the direct purchase of materiel and/or support and services, the insurgent organization can obtain logistics in ways to include the following:

- Theft from the governing authority.
- Voluntary donations from a supportive relevant population.
- Tax and/or levy on local and area populations.
- Specified materiel from a relevant population in exchange for social services and/or health welfare support.
- Self-manufacturing of designated materiel.

Communications and Tradecraft Cell

2-135. The communications and tradecraft cell is the communications planner for the insurgent organization. The cell advises insurgent organization leaders on the feasibility of all insurgent activities from a communications perspective. It also determines the internal communications equipment necessary for the success of direct action missions.

2-136. This cell facilitates communications inside and outside of the organization with a courier service, dead-drop locations, and other tradecraft (clandestine) communications. It also provides multiple electronic means for digital and secure communications. It is equipped with computers, radios, small satellite communication antennas, and long-range cordless telephones. The cell is a source of communications, Internet, electronic tradecraft, and steganography expertise and provides training as required. It maintains close coordination with the INFOWAR cell for Internet communications.

2-137. This cell provides courier service as s simple and secure means of communication among cells in the insurgent organization. Transportation for a courier may be a civilian model motorcycle, moped, motor scooter, bicycle, sedan, or even a taxi. The courier probably will not have an overt weapon or radio and will appear to be a noncombatant. Another insurgent may ride with the courier as a lookout or to provide security. Depending on the circumstances, some couriers may not use vehicles and will be used instead as foot messengers. Couriers may be male, female, or even children. Messages may be written, encrypted, or memorized and presented orally in person.

Finance Cell

2-138. The finance cell is the financial and economic planner and provider for the insurgent organization. It determines the internal financial requirements necessary for the success of direct action missions. This cell plans local fund-raising activities such as bribery, extortion, robbery, and operating front companies in order to finance and resource the insurgent organization's activities. It also has links to regional, national, and international fund-raising activities. The cell also determines the roles of direct action cells and other cells in acquiring additional funds for the insurgent organization.

2-139. Finance personnel establish and monitor internal and external funding and funds management mechanisms. Skilled operators appeal to the local and international community, possibly through the media, for political, monetary, and logistics support. They may do this in coordination with the INFOWAR cell and logistics cell.

2-140. Voluntary or coerced contributions from the diaspora of a relevant population are a significant enterprise of regional, international, and transnational connections for the insurgency. Both governmental and nongovernmental donations and grants may wittingly or unwittingly finance an insurgency.

Shelter Cell

2-141. The shelter cell plans for and provides secure accommodations at safe houses for direct action and supporting cells. Safe houses may also accommodate visiting and high-ranking insurgents. These shelters are closely coordinated with the intelligence, planning, logistics, and transportation cells. Caches of materiel can be aligned with locations of safe houses and/or transit routes to and from insurgent areas of interest and territorial safe havens.

2-142. Sympathizers in the relevant population may volunteer their homes, equipment, vehicles, and services. Maximum use is made of local assistance and facilities regardless of capability.

Training Cell

2-143. The training cell plans and coordinates training for the insurgent organization members. Training covers the basic organization, duties, and responsibilities of the insurgent organization, and advances to specialized training of insurgents for cell or team responsibilities.

2-144. Basic instruction includes—
- Basic marksmanship.
- Surveillance and intelligence-collection methods.
- Basic tactical instruction.
- Communications techniques.

2-145. Examples of specialized instruction may include advanced tactical instruction such as conduct of—
- Ambushes.
- Assaults.
- Raids.
- Assassination.
- Bombing with IEDs.
- Sniper activities.
- Advanced surveillance and intelligence-collection methods.
- Kidnapping.
- Extortion.

2-146. Training may be centralized or be conducted at the cell level depending on the structure and mission of the insurgent organization. Generally, training is a combination of both with the basics taught in a centralized location and specific mission and functional training conducted at the local cell level. Training may be distributed via the Internet. Training actions are coordinated closely with the planning and recruiting cells.

Recruiting Cell

2-147. The recruiting cell provides the manpower resourcing and recruitment planning and for integrating recruits into the insurgent organization. Recruiting can promote radical religious, militant ethnic, nationalist, or social agendas that propose to remedy compelling grievances within a relevant population. Recruits might not be aware of the true nature of the organization they are joining. Recruiting may be wittingly or unwittingly financed from both governmental and nongovernmental donations and grants.

2-148. Often, legitimate organizations can serve as recruiting grounds for insurgent organizations. The organizations from which individuals can be recruited need not necessarily be violent or illegal themselves, but simply contain populations that are sympathetic to the same goals as the insurgent organization. Recruiting may be for particular skills, training, and/or qualifications and may not be tied to ideological characteristics. Insurgent organizations may attempt to recruit current and former members of national armed forces, both as trained operatives and as agents in place.

2-149. The recruiting cell uses many varied and different methods to persuade potential insurgents to join them. Some of these persuading factors may be monetary, religious, ethic, nationalistic, anger, promise of power, or fear. The Internet is a powerful recruitment tool. The recruiting cell maintains close coordination with the INFOWAR cell.

2-150. Insurgents may also use coercion and leverage to gain limited or one-time cooperation from useful individuals. This cooperation can range anywhere from gaining information to conducting a suicide bombing. Blackmail and intimidation are common forms of coercion. Threats to family members are also employed. Coercion is often directed at personnel in government security and intelligence organizations.

2-151. Internal security is the primary concern of the insurgent organization when recruiting. First, a potential recruit must pass an intense screening process. Once selected for recruitment, he/she is then closely monitored by the CI and internal security cell prior to full recruitment and integration into the insurgent organization.

Transportation Cell

2-152. The transportation cell plans and facilitates transportation for the insurgent organization. It responds to transportation requirements of other cells in the organization, especially the logistics and direct action cells. If sufficient or specific transportation is not available, the transportation cell either provides or arranges for it.

2-153. Insurgents may have no vehicles or supplies at all and depend completely on caches, porters, or other transportation or supply means. Local sympathizers may volunteer their equipment and services. Depending on the mission, the local insurgent organization may be augmented by any and every type of personnel and/or vehicle. Insurgents may requisition or confiscate local civilian transportation assets and materiel. This includes the use of civilian personnel for drivers, porters, lookouts, and security personnel. Draft animals may also be used as bearers and/or porters.

2-154. The vehicles in the transportation cell are indistinguishable from civilian vehicles and are always kept as dispersed as possible, in order to prevent detection and destruction by the governing authority. Rarely, if ever, will all vehicles in the cell be colocated. Whenever possible, vehicles are dispersed for use by locals as commercial, delivery, agricultural, general cargo, construction, and general-purpose vehicles used in everyday life. When required, the transportation cell assembles the appropriate mix of vehicles to transport items and/or personnel to a specific location. After performing the necessary transportation tasks, the vehicles then melt back into the general population and environment.

Civil Affairs Cell

2-155. The civil affairs cell is responsible for the planning, preparation, and implementation of all civil affairs activities for the insurgent organization. It organizes a synchronized program of actions for the benefit of a relevant population. The cell may work openly and/or discretely with the INFOWAR cell for media manipulation to ensure the insurgent organization gets credit for providing those benefits.

2-156. Key civil affairs events are digitally recorded on digital video and still cameras in order to publicize results and successes intended to create a positive image in the relevant population. Some activities may be staged to enhances the prestige of the insurgent organization and/or present a negative image of the governing authority. The recordings are transferred to the INFOWAR cell for manipulation and exploitation and/or released to sympathetic media for local or worldwide distribution.

Medical Cell

2-157. Insurgent medics are combatants. When necessary, they fight. Insurgent medical personnel may be a mixture of men and women. Women may make up 50 percent or more of the medical cell strength.

2-158. The insurgent organization has limited medical capability. However, insurgent medical care is coupled with local medical assets in the area whenever possible. Maximum use is made of local medical assistance and facilities regardless of medical capability.

2-159. Insurgents will persuade and/or coerce local civilian medical support when needed. Sympathizers in the local populace may volunteer their homes, equipment, vehicles, and services for ad hoc medical care. Local medical personnel may volunteer to treat the ill and wounded. Sympathizers may also assist in the evacuation of wounded insurgents to civilian, militia, state, or even military facilities. Evacuation means can include general-purpose cargo vehicles, carts, or even taxis.

2-160. When necessary, supported direct action cells receive litters from the medical cell to transport wounded. The supported cell provides its own litter bearers. Noncombatants may also be forced to serve as litter bearers.

2-161. A medical aid station is usually set up in a fairly safe area, while other medics may accompany direct action cells or other insurgents. When necessary, medical functions are performed in tents, tunnels, caves, or local accommodations. In some cases, the medical cell may colocate with a village clinic. Insurgents may or may not have the services of a civilian medical officer (physician). If available, the physician can provide immediate trauma stabilization and minor surgical intervention. Meanwhile, the medics provide limited medical intervention, minor surgery, and treatment of most common illnesses and lesser wounds.

2-162. Severe and long-term medical care relies on evacuation to civilian or other medical facilities. More routine and excess ill and wounded are transported to civilian medical facilities or may be cared for in insurgent safe havens.

2-163. As the insurgent organization establishes gradual control in designated areas, medical care extends beyond the role of preserving capabilities of only the insurgent organization. Basic preventive medicine and medical care can be offered to a relevant population as a means to encourage and develop popular support.

This page intentionally left blank.

Chapter 3

Guerrillas

This chapter presents an overview of guerrilla organizations and actions as part of the irregular OPFOR for training U.S. forces. Training conditions presented by this type of OPFOR are a composite of real-world guerrilla forces and indicate guerrilla capabilities and limitations that may be present in actual operational environments (OEs). Guerrilla combat power can be enhanced by possible affiliations with other combatants such as insurgents, criminal elements, special-purpose forces (SPF), or regular military forces. Passive or active civilian supporters can expand guerrilla capabilities.

GENERAL CHARACTERISTICS

3-1. A *guerrilla force* is a group of irregular, predominantly indigenous personnel organized along military lines to conduct military and paramilitary operations in enemy-held, hostile, or denied territory (JP 3-05). Thus, guerrilla units are an irregular force, but structured similar to regular military forces. They resemble military forces in their command and control (C2) and can use military-like tactics and techniques. Guerrillas normally operate in areas occupied by an enemy or where a hostile actor threatens their intended purpose and objectives. Therefore, guerrilla units adapt to circumstances and available resources in order to sustain or improve their combat power. Guerrillas do not necessarily comply with international law or conventions on the conduct of armed conflict between and among declared belligerents.

SCOPE AND DURATION OF OPERATIONS

3-2. The area of operations (AOR) for guerrilla units may be quite large in relation to the size of the force. The reason for this is that a large number of small guerrilla units can be widely dispersed. Guerrilla operations may occur as independent squad or team actions. In other cases, operations could involve a guerrilla brigade and/or independent units at battalion, company, and platoon levels. A guerrilla unit can be an independent paramilitary organization and/or a military-like component of an insurgency. Guerrilla actions focus on the tactical level of conflict and its operational impacts. Guerrilla units can operate at various levels of local, regional, or international reach. In some cases, transnational affiliations can provide significant support to guerrilla operations.

3-3. Guerrilla forces are adaptive, flexible, and agile in quickly changing their composition to optimize organizational capabilities against known or perceived vulnerabilities of an enemy. Guerrillas exploit familiarity with their physical environment and the ability to blend into the local populace. Small guerrilla units have great mobility and ability to move throughout enemy-occupied areas.

3-4. Guerrillas seek to gain small psychological victories. These victories do not need to be significant in terms of material damage to the enemy. These tactical victories only need to show that a small guerrilla force can defeat [at least parts of] a much larger enemy force.

3-5. Guerrilla forces take prudent risks when an expectation exists for successful attack on an enemy, but may also make significant practical sacrifices in individuals and materiel in order to achieve a major psychological impact on an enemy. Guerrillas also apply information warfare (INFOWAR) capabilities to weaken or exhaust enemy resolve.

3-6. Ultimately, the resolve of guerrilla leaders and members of guerrilla organizations determines how long to continue guerrilla operations. Time is a key factor that guerrilla forces use as a combat multiplier in a long-term commitment to degrade and eventually defeat the will of an enemy. The goal is not necessarily to defeat enemy forces but to outlast them. This long-term struggle includes a full range of actions that range from espionage and media manipulation to more violent actions such as sabotage, assassination, bombing, ambushes, and raids. Guerrillas can use acts of terrorism to achieve either selective or random psychological stress and physical damage or destruction. Actions are typically quick and violent, followed by rapid dispersal of assembled guerrilla forces.

3-7. Factors that affect the scope and duration of guerrilla operations include—
- The level of sympathetic support from an indigenous relevant population.
- Regional sociological demographics and ethnic-racial relationships and tensions.
- Governance actions and support services to a relevant population by a civil or military authority in a contested region.
- Economic stability and disproportionate distribution of benefits to a relevant population.
- Physical and seasonal aspects of topography and climate.
- The amount of covert or overt support from organizations or states with interests in a contested region.

Note. TC 7-100.2 complements this publication in regard to tactics, and FM 7-100.4 provides details of organization, manning, weapons, and equipment. For weapons and equipment data, see the *Worldwide Equipment Guide*.

RELATIONSHIPS WITH OTHER ORGANIZATIONS AND ACTORS

3-8. Guerrillas can act separately from other groups, organizations, and/or activities in conflict with the same enemy or act in conjunction with them to pursue common objectives. Depending on local or regional conditions, some guerrilla units may be affiliated with or subordinate to regular forces or an insurgent organization, or they may operate as independent of such organizations. Regular forces can provide overt and covert support for guerrilla operations, including the expertise of advisors, liaison teams, and SPF. Guerrilla units can be associated with regular forces on a temporary basis for particular missions, but would be incorporated into regular forces only when their capabilities are similar to those of the regular forces operating in the same geographic area.

Note. Affiliated organizations are those operating in another organization's AOR that the latter organization may be able to sufficiently influence to act in concert with it for a limited time. No command relationship exists between an affiliated organization and the organization in whose AOR it operates.

3-9. Guerrillas are more likely to be incorporated in and subordinate to an insurgent organization when both are parts of the irregular OPFOR. A guerrilla organization that is affiliated with or subordinate to an insurgent organization may also be affiliated with SPF, other regular military forces, and/or criminal elements. Guerrillas will generally accept help from any other organization as long as it meets a need and is compatible with the guerrilla force's interests and objectives.

3-10. Guerrillas can be part of the Hybrid Threat (HT). The HT can be any combination of two or more of the following components: regular forces, irregular forces (such as guerrillas and/or insurgents), and/or criminal elements. Possible HT combinations include guerrillas operating openly with regular military forces or various forms of covert cooperation or support. (See TC 7-100 for detailed discussion on the HT.)

3-11. Guerrilla associations with other actors in a contested region can vary, and relationships may change periodically during a long-term conflict. Guerrillas may have relationships with local or higher-level insurgent organizations or criminal organizations. If a standing relationship does exist among guerrillas and other irregular OPFOR or HT actors, any allegiance or affiliation may be focused on single-issue agreements or a mutual ideological commitment. Affiliation with criminal organizations is dependent upon the mutual needs of a criminal organization and a guerrilla unit or may be a contractual arrangement for specific tasks. Guerrilla activities often overlap with criminal activities.

3-12. In addition to possible affiliations with insurgent, criminal, or regular military organizations, guerrilla units may also be affiliated with supportive civilians, perhaps covert informal support networks acting with a façade such as a charitable organization. The support provided by different categories of civilians can include—
- Coerced support of guerrilla actions.
- Passive support by people sympathetic to the goals of the guerrillas.
- Actors actively supporting and engaging in combat support or sustainment of the guerrillas.

Guerrillas may depend heavily on the active and passive support from the local population (see chapter 1).

Note. Active sympathizers may provide important logistics services but not directly participate in combat operations. If they participate in guerrilla activities, they become guerrillas.

AFFILIATION WITH SPECIAL-PURPOSE FORCES

3-13. When guerrilla organizations are affiliated with the military forces of an external nation-state, they are most likely to be associated with special-purpose forces (SPF). SPF carry out operations either independently or in coordination with regular and/or irregular forces (such as guerrillas). In some cases, SPF can provide funding for specific guerrilla activities or for the entire guerrilla organization. (See chapter 15 of TC 7-100.2 for more information on SPF.)

3-14. The nature of shared goals or interests determines the tenure and type of relationship and the degree of affiliation between SPF and guerrillas. For example, the affiliation of an SPF detachment with guerrilla organizations is dependent only on the needs of the guerrilla organization or on the needs of the SPF at a particular time. The relational dynamics of SPF units are very fluid and apt to change from one day to the next. For example, if an SPF mission requires close cooperation with guerrilla forces, those guerrilla units can be included in the SPF task organization with the appropriate command and support relationships.

3-15. SPF can recruit, organize, train, advise, and support guerrillas and conduct (or lead) operations in conjunction with them. SPF personnel may fight alongside such affiliates or assist them to prepare for offensive actions, diversionary measures, or other missions. In some cases, the SPF will not only advise and assist but actually control (command) the guerrilla units as a surrogate force. When guerrilla forces support SPF teams, the SPF teams can serve as the planning and command element for these forces.

3-16. SPF sappers can accompany and/or augment guerrilla units. SPF sappers can train affiliated guerrillas as sappers. SPF battalions using sappers in an assault and/or demolition role may need to form several sapper platoons. In that case, the additional sapper platoons may be manned by affiliated guerrillas or may be a mixture of SPF and guerrillas. SPF sappers can guide guerrillas through enemy lines and obstacles to perform their missions (or guerrilla sappers can do the same for SPF teams). An SPF sapper team may serve with and train guerrillas on how to infiltrate, set demolitions, and assault enemy installations. In some cases, the SPF sappers may train guerillas in the manufacture and employment of improvised explosive devices (IEDs) and signal flares. In other cases, an SPF sapper team may manufacture the IEDs and signal flares and then give them to guerrillas to emplace and/or detonate.

3-17. The deep attack and reconnaissance platoon of an SPF battalion typically operates in enemy-held territory. Its missions can include assisting guerrillas in offensive actions or providing communications, liaison, and support to stay-behind guerrillas and guerrilla activities in the defense. Almost any type of SPF unit can do the same.

3-18. The long-range signal platoon of an SPF battalion or a signal team from an SPF company can assist in training affiliated guerrillas on how to set up, operate, maintain, and transport communications equipment. SPF signal teams may accompany and/or augment guerrilla units and support guerrilla operations. A single small SPF signal team can provide long-range communications support for guerrilla units up to battalion size. A full SPF signal team can do the same for a brigade-size unit. This team may also serve in a signals reconnaissance collection role.

3-19. Various other types of SPF teams may accompany and/or augment guerrilla units and provide them with support whenever necessary. Such teams include the following:

- SPF sniper teams, who may serve as part of a stay-behind element or a hunter-killer (HK) team. SPF snipers can also train guerrillas to serve as snipers or marksmen.
- SPF mortar teams, who may provide mortar support to guerrilla forces. They can also provide mortars to the guerillas and train them in the proper use and maintenance of mortars.
- SPF air defense teams, who may provide air defense support to guerrilla forces. They can also provide air defense weapons to the guerillas and train them in the proper use and maintenance of those weapons.
- SPF medical teams, who may provide medical supplies and medical support to guerrilla forces. Each SPF medical team is designed to provide medical support for guerrilla units up to brigade size. A small medical team can support a battalion-size force. These teams can also train guerrilla forces on how to perform emergency medicine, battlefield medical procedures, and evacuation.
- SPF unmanned aerial vehicle (UAV) teams, who may provide UAV support to guerrilla forces, including the acquisition of reconnaissance information on targets or facilities.
- SPF diver teams, who may provide diving support to guerrilla forces, including water infiltration, reconnaissance, and demolition. These teams can also provide diving equipment to guerrilla forces and train them in its use.

Note. In some instances, SPF can provide guerrilla units with new or high-technology niche weapons and equipment and train them to use it.

3-20. When SPF are to be air-dropped into an area known to contain a guerrilla force of sufficient size and nature to warrant cultivating as an affiliated force, the guerrilla force, in all probability, will be receptive to outside support. Other SPF or regular military forces may have trained the guerrilla force as drop zone reception personnel. Once on the ground, the SPF team or detachment attempts to make contact with the guerrilla forces.

Logistics

3-21. Because guerrillas typically operate in enemy-held, hostile, or denied territory, they must acquire a self-sustaining capability. Although they typically rely on some support from the local population, they need to avoid enemy retribution on the population and/or dissatisfaction of the population with the demands of supporting a guerrilla force. One example of self-sustainment is to acquire weapons, ammunition, supplies, transportation assets, and other commodities by raiding or ambushing enemy forces, locations, and installations. In another example of self-sustainment, civilian medical facilities may be used to treat guerrillas on a case-by-case clandestine basis, but a preferred capability is to develop medical treatment stations and convalescence sites integral to guerrilla safe havens or complex battle positions.

3-22. Guerrilla units normally have sufficient assets to transport munitions and materiel for the immediate short-term fight. Organic transportation and use of available resources of the local area provide the guerrillas a degree of autonomy and freedom of action. For a sustained fight, however, guerrillas normally require additional support and sources to transport or stage materiel. Such sources can include transport assets from higher guerrilla organizations to transport materiel or preposition items in caches for resupply. For example, a guerrilla battalion might depend on support from the brigade's transport company, or a guerrilla company might receive support from the battalion's transport platoon. Other external sources may include civilian augmentation.

3-23. A guerrilla brigade or its subordinates may be augmented by military or civilian vehicles (motorcycles, trucks, cars, bicycles, carts, or high-mobility/all terrain vehicles) or personnel depending on the mission. Local sympathizers may volunteer their equipment and services. The guerrilla commander may requisition or confiscate local civilian transportation assets and materiel. This includes the use of civilian personnel for porters. Vehicles are a mix of military and civilian. The guerrillas may have no vehicles at all and depend completely on caches, porters, or other transportation or supply means.

3-24. Requirements for obtaining supplies and services from a local population are evaluated in conjunction with the sustainment needs of the population. Support to guerrillas from communities is coordinated with a willing population, but is sometimes coerced due to tactical needs of the guerrilla force. Guerrillas recognize that recurring requisition or confiscation of materiel may alienate the population. Therefore, the guerrillas may institute a system for receipts and eventual compensation of the citizenry.

3-25. Sustained combat and/or participation in a fight by guerrilla units may require significant operational, administrative, and logistics support from regular military forces operating in the same geographic area. This type of regular forces support complements the significant sustainment that guerrillas require from a local or regional populace. As a basic level of logistics support, guerrillas subsist primarily from locally available resources. This limitation influences the size of a guerrilla unit and its proximity to other guerrilla units.

PERSONNEL

3-26. Guerrillas often come from impoverished backgrounds and are used to hardships. Typically, they are young and in good physical condition. They can make do with less by both design and background.

3-27. The military-like organizational structure and intent of a guerrilla unit is often accompanied by a military style of clothing or uniform, equipment and weapons, and distinctive badges or insignia. However, a guerrilla may not always wear a military-like uniform. They may wear civilian clothing and be indistinguishable from the local population (other than weaponry, which they may conceal or discard). Although most guerrillas are armed, they may seek to avoid being identified as guerrillas by giving the appearance that their weapons are part of their normal civilian occupation or activities. Weapons and other materiel may be concealed until they are needed for a guerrilla action.

3-28. Some, or all, guerrillas may be part-time fighters and melt back into the civilian population when not assembled to conduct operations. Guerrillas may continue their normal positions in society and lead clandestine lives as members of a guerrilla organization. They may be a local resident one moment and a guerrilla the next.

3-29. Guerrillas may be a mixture of men, women, and children. Women and children may be used as runners, messengers, scouts, guides, suicide bombers, drivers, porters, snipers, lookouts, or in other roles. They may also emplace and/or detonate IEDs, signal flares, and mines. Women (and possibly children) may be fighters and participate in drive-by shootings, assassinations, and/or assaults.

3-30. Since most guerrillas are indigenous to the area in which they are fighting, their knowledge of the local population, customs, issues, language, and terrain are first hand. They use this understanding to develop working relationships with the populace. They can apply this knowledge to win support, develop intelligence networks, gain new recruits, and/or develop effective propaganda. If the guerrillas are not able to persuade the local populace to provide support, they have the force to coerce them.

> *Note.* Members of a guerilla force are predominantly indigenous to the region but may include personnel who are not indigenous to a particular guerrilla unit's AOR and may have characteristics significantly different from most of the guerrilla force or the local population.

BASE CAMPS AND TRAINING CAMPS

3-31. Like any other armed force, guerrillas have requirements for C2, rest, resupply, refit, and training. The larger the guerrilla force and the more active they are, the more they will need established bases, both semipermanent and temporary.

3-32. Guerrillas who are part-time fighters may prefer to continue to live in their own homes. They would assemble into guerrilla units only for the time necessary to carry out a mission and then return to their homes. They would establish and occupy base camps only if security does not permit them to live at home or in order to prepare for actions involving more than just a few squads or HK teams. Platoon and larger organizations typically operate out of a base camp.

3-33. Guerrilla forces may need to avoid large concentrations of troops in camps. Even though logistics conditions may permit large troop concentrations, commands normally are broken up into small camps and widely dispersed. Dispersion facilitates concealment, mobility, and security. Large forces may be concentrated to perform specific operations. On completing the operation, the force is dispersed quickly into small units.

3-34. Base camps must be relatively safe and secure in areas where guerrillas can rest, eat, and plan. More sophisticated guerrilla base camps have command posts, training areas, communications facilities, medical stations, and logistics centers. These base camps, however, are not the same as bases of regular military forces. They are usually small in overall scope, spread out, and sometimes underground. Guerrillas try to locate base camps within guerrilla-controlled areas where cover and concealment provide security against detection. In rural areas, base camps tend to be in areas where the guerrillas have popular support. Urban guerrillas may rent or confiscate houses for use as temporary bases for small guerrilla forces, such as HK teams. Guerrillas will normally avoid battling over their base camp. If the site is discovered, the guerrillas will abandon it and move to another location. Routes into a base camp will be constantly observed for security. Mines, signal flares, IEDs, and ambushes are used as standard security enhancements. If surprised or cornered, guerrillas will vigorously defend themselves with a delaying action while evacuating key personnel and equipment.

3-35. Common characteristics of guerrilla base camps are—
- Covered and concealed.
- Located on remote, rough, and inaccessible terrain.
- Suitable for bivouac.
- Well defined and defended perimeter.
- Planned escape routes.

3-36. Guerrilla base camps are temporary or semipermanent and depend on secrecy for their existence. These bases are kept small, and usually there is more than one base in the guerrilla unit's AOR. Guerrillas typically use four types of base camps:
- The main camp.
- Reserve camps that are in a preselected location and would be occupied if the main camp must be permanently evacuated.
- Temporary camps located close to current combat operations and occupied only during those operations.
- False camps located away from main and reserve camps to deceive enemy forces.

3-37. Guerrillas favor level, well-drained campsites with good water supply, natural fuel, cover, and adequate vegetation to provide concealment from visual observation. The preferred camps are also chosen with a view toward easy access to the target population, access to a friendly or neutral border, good escape routes, and good observation of approach routes for enemy forces. When enemy operations force the guerrillas out of their preferred base camps, they tend to establish camps in rugged, inhospitable areas not easily penetrated by enemy forces.

Base Camp Security

3-38. To provide security against detection, guerrillas attempt to locate base camps in areas with cover and concealment. Usually, the rougher the terrain, the less likely is the chance of being surprised by enemy forces. While guerrillas avoid defensive combat, they emphasize short-term defensive action in the base camp vicinity to aid evacuation, if necessary.

3-39. Base camps are usually in relatively remote areas for security. To preclude accidental discovery, base camps are not usually near inhabited areas. However, because the guerrillas must be able to fulfill their logistics needs, their base camp usually will not be more than a day's march from a village or town.

3-40. Guerrilla camps are normally located away from roads, main trails, and civilian communities. The surrounding terrain should obstruct rapid enemy approach into the area. The site of the camp should provide concealed evacuation routes. The camp is surrounded by guard and warning systems. Prepared positions are organized to delay or destroy enemy forces when this action becomes necessary or desirable. Almost every base camp is surrounded by observation posts, preferably located on hilltops. From these posts, an early alert is passed enabling timely evacuation of the area.

Training Camps

3-41. Guerrilla training camps may be established in both urban and rural environments. While some training is accomplished in urban safe houses or rural guerrilla base camps, most training is accomplished at locations focused on training. This may be a special urban safe house, a remote guerilla base camp, or a training center in another country.

GUERRILLA ORGANIZATIONS

3-42. Guerrillas use a military-like organizational structure for C2 and conduct of operations. For example, the basic building block of a guerrilla organization may be a squad consisting of two fire teams. Such squads are the basis for building guerrilla platoons, companies, battalions, and brigades. However, guerrilla commanders can task-organize these units for specific actions. Even prior to specific actions, whole guerrilla companies may already be restructured (task-organized) as hunter-killer (HK) companies, made up HK groups, HK sections and HK teams. When a guerrilla battalion consists predominantly of HK companies, it may be called a guerrilla HK battalion. When a guerrilla brigade consists predominantly of HK battalions (or conceivably of multiple separate HK companies), it may be called a guerrilla HK brigade.

> *Note.* Although this chapter and FM 7-100.4 present typical structures of guerrilla units, these forces are also irregular in the sense that individual units can vary in manning, weapons, and materiel.

3-43. Guerrilla organizations may be as large as several brigades or as small as a platoon and/or independent HK teams. Often a brigade-size guerrilla force may not be appropriate for a particular mission or area AOR. It may be too large, and a task-organized guerrilla battalion may be sufficient. An example task-organized battalion might have four or five HK companies, organic battalion units, a weapons battery (with a composite of mortar, rocket launcher, and antitank platoons) from brigade, and possibly intelligence and INFOWAR augmentations.

3-44. The hierarchy of military-like terms for guerrilla units (from the bottom up) is as follows:
- Team or HK team.
- Squad or HK section.
- Platoon or HK group.
- Company or HK company.
- Battalion (or HK battalion).
- Brigade (or HK brigade).

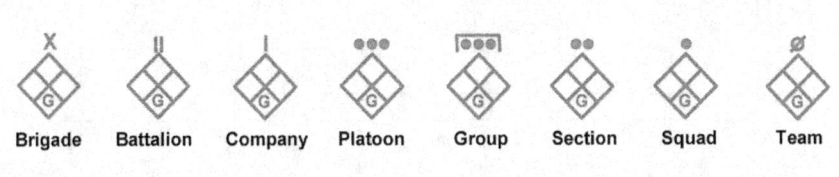

Figure 3-1. Guerrilla organization symbols: brigade to team level

> *Note.* Some guerrilla organizations may have honorific titles that do not reflect their true nature or size. For example, a guerrilla force that is actually of no more than battalion size may call itself a "brigade," a "corps," or an "army." A guerrilla organization may also refer to itself as a "militia." This is a loose usage of the term *militia*, which generally refers to citizens trained as soldiers (as opposed to professional soldiers), but applies more specifically to a state-sponsored militia that is part of the state's armed forces but subject to call only in emergency. To avoid confusion, the TC 7-100 series uses *militia* only in the latter sense.

3-45. Guerrilla units can be independent units or can be associated with insurgent organizations at local, regional, provincial, national, or transnational levels. Organizational relationships between guerrillas and insurgents can fluctuate and be mission dependent, event-oriented, mutually coordinated, and/or coerced for a specific temporary purpose. The relationship may be one of loose affiliation or involve a more formal command relationship. When an insurgency includes guerrilla units, the units may be under the C2 of a local insurgent organization or of a higher insurgent organization. The relationships may be temporary and remain in effect only as long as the both insurgent organization and guerrilla unit mutually benefit.

> *Note.* The guerrilla organization diagrams of FM 7-100.4 (on which examples in this chapter are based) and the personnel and equipment lists that accompany them are a baseline that U.S. Army trainers can modify to provide the appropriate level of combat power as part of the conditions required for a particular training exercise and/or training task. The organizational directories of FM 7-100.4 provide detailed information on manning, weapons, and equipment for a typical guerrilla brigade and its subordinate units, which represent a composite of actual guerrilla forces.
>
> FM 7-100.4 provides detailed step-by-step instructions on how to construct a task organization based on the training requirements. FM 7-100.4 also describes how to select equipment options. See also, TC 7-101 for guidance on creating the appropriate OPFOR order of battle during exercise design.

GUERRILLA BRIGADE

The composition of a guerrilla brigade may vary. The structure depends on several factors, including the physical environment, sociological demographics and relationships, economics, and support available from external organizations and/or countries. A guerrilla brigade may have over 4,000 guerrillas or may have specified units significantly reduced in strength. Figure 3-2 shows an example of such a brigade.

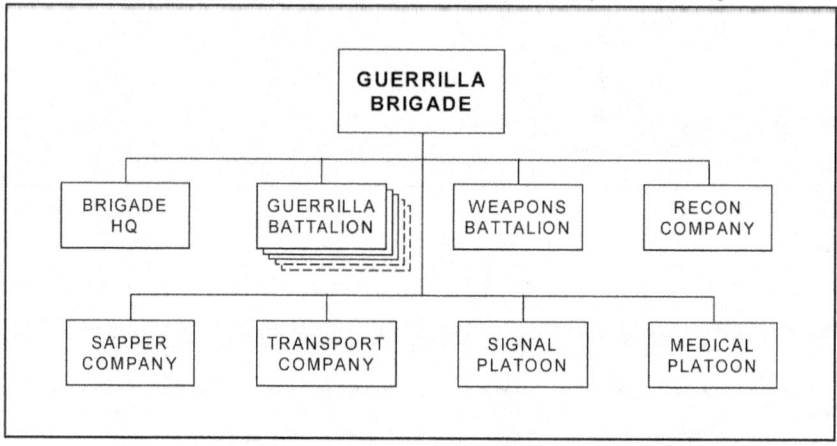

Figure 3-2. Guerrilla brigade (example)

3-46. The typical composition of a full-strength guerrilla brigade may be as follows:
- Brigade headquarters.
- Three or more guerrilla battalions.
- Weapons battalion.
- Reconnaissance company.
- Sapper company.
- Transport company.
- Signal platoon.
- Medical platoon.

3-47. Guerrilla units are tailored organizational structures based on the conditions of a particular OE and available manning, weapons, and equipment. The following are some examples of such tailoring:
- In a particular OE, a guerrilla brigade could be a multi-battalion combat force operating in several AORs, with a full range of combat support (CS) and combat service support (CSS) units (as in figure 3-2) and possibly affiliated with regular military forces.
- In mountainous or forested areas with no major population centers, a guerrilla brigade might have only one or two battalions (or five or six separate companies) with little or no additional CS or CSS. Options could include widely dispersed companies depending on sustainment from local resources in a rural context or materiel obtained from defeated enemy forces and isolated enemy installations.
- A guerrilla brigade based in a major urban-industrial center or a megalopolis would probably have a distinctly different structure of combat, CS, and CSS capabilities, tailored to readily available resources within a densely packed population and urban terrain.

GUERRILLA BRIGADE HEADQUARTERS

3-48. The guerrilla brigade headquarters consists of the brigade command group, brigade staff, and a headquarters and service section. The personnel and functions of these organizations are described below.

Brigade Command Group

3-49. The brigade command group includes the brigade commander and deputy commander (DC). Guerrilla commanders have complete authority over their subordinates and overall responsibility for those subordinates' actions. In the event the commander is killed or incapacitated, the DC assumes command. The brigade chief of staff (COS) is considered part of the command group but actually heads the brigade staff.

Brigade Staff

3-50. The brigade COS exercises direct control over the brigade staff and is in charge of the main command post (CP) in the absence of the brigade commander. His role is to direct staff planning and coordinate all staff inputs that assist the commander's decisionmaking. He is the commander's and DC's focal point for knowledge about the friendly and enemy situation.

3-51. The brigade staff has three primary staff officers: operations officer, intelligence officer, and officer. Each of these heads one the following staff sections:
- Operations section.
- Intelligence and information section.
- Resources section.

Operations Section

3-52. The brigade operations section is the basis of the main CP. The brigade COS normally resides with this section. Operational signature is small. Whenever possible, staff functions are performed in tents, tunnels, caves, or local accommodations. When mounted, the main CP usually operates from command and staff vehicles. The brigade commander determines how to organize C2 of the brigade and how many CPs will be activated. The operations section is structured to simultaneously support a main CP and an auxiliary CP. A forward CP may also be required.

Note. Guerrilla brigades and battalions may not have all the types of CPs discussed here. However, specific situations may require various types.

3-53. The brigade operations section includes the operations officer (who also serves as the deputy COS), the chief of current operations (who also serves as assistant operations officer), and the chief of future operations. The latter two officers each head subsections.

3-54. **Current Operations Subsection.** The chief of current operations serves as the representative of the commander, COS, or operations officer in their absence and has the authority to control forces in accordance with the commander's intent.

3-55. **Future Operations Subsection.** The chief of future operations monitors the friendly and enemy situations and their implications for future actions. He advised the commander on how to make adjustments to the battle plan during the fight.

3-56. **Functional Staff.** Functional staff elements may augment the brigade operations section when required. These experts in particular functions (such as fire support) advise the brigade command group and staff on issues pertaining to their individual areas of expertise.

3-57. **Liaison Teams.** The operations section may also receive liaison teams from subordinate and affiliated units that perform tasks in support of those. Liaison teams are not a permanent part of the brigade staff structure. They support the brigade staff with detailed expertise in particular functional areas and provide direct communications to their parent organizations. SPF advisors and liaison personnel from regular military forces may operate from the brigade headquarters or may be located at lower echelons. The brigade staff may also receive liaison and advisory teams from organizations covertly or overtly supporting guerrilla operations. Their purpose may be based on functional military expertise for guerrilla unit operations or may support selected political, economic, social, or INFOWAR tasks. If the guerrilla brigade is affiliated with or subordinate to an insurgent organization, a political advisor may also be present.

Intelligence and Information Section

3-58. The intelligence and information section provides intelligence on the current and future OE. It also provides insights on opportunities for ongoing and future operations and windows of opportunity. At least one representative from the intelligence and information section is part of, and accompanies, each auxiliary or forward CP. This section has three subsections: reconnaissance, INFOWAR, and communications.

3-59. **Reconnaissance Subsection.** The chief of reconnaissance, who heads this subsection, develops reconnaissance plans, gathers information, and evaluates data on the OE. He supervises the efforts of subordinate reconnaissance units.

3-60. **INFOWAR Subsection.** The chief of INFOWAR, who heads this subsection, coordinates the employment of INFOWAR assets, both at brigade level and in subordinate units. This subsection oversees the planning and execution of all elements of INFOWAR (see appendix A).

3-61. **Communications Subsection.** The chief of communications and his subsection plan the use of all forms of communication and communications security. The subsection may be augmented by personnel from the brigade's signal platoon when necessary.

Resources Section

3-62. The resources officer is responsible for the acquisition, distribution, and care of all the brigade's resources, both human and materiel. He is also in charge of the brigade's sustainment CP and may establish multiple sustainment CPs. The resources section may be augmented from the brigade's transport company when necessary. The section is also structured to accommodate augmentation from the functional staff and liaison teams. It consists of two subsections: logistics and administration.

3-63. **Logistics Subsection.** The chief of logistics heads the brigade logistics system. He supervises transportation support. He is responsible for the acquisition and distribution of supplies to sustain the brigade. He is also responsible for the all armament accountability, readiness, supply, utilization, repair, and evacuation.

3-64. **Administration Subsection.** The chief of administration supervises all personnel actions and transactions. This subsection assigns personnel and records losses.

Headquarters and Service Section

3-65. The headquarters and service section provides supply and transportation support to the brigade headquarters. It includes a command team, a support team, and a supply and transport team.

WEAPONS BATTALION

3-66. The weapons battalion of a guerrilla brigade typically consists of the following units:
- Battalion headquarters.
- Towed mortar battery.
- Rocket launcher battery.
- Antitank battery.
- Signal section.
- Transport section.

This battalion provides fire support to any other subordinate units of the guerrilla brigade as necessary. In some instances, due to the dispersed nature of guerrilla operations, a battery or parts thereof may be allocated to a particular guerrilla battalion.

Battalion Headquarters

3-67. The battalion headquarters consists of a command section and a staff section. The weapons battalion commander locates where he can best support the brigade operations and is often positioned in the vicinity of the brigade commander. The command section also includes the battalion's deputy commander, who would be in charge of a forward or auxiliary CP, if necessary. The chief of staff (COS) is part of the command group and staff. However, he is usually located with the staff on the battlefield, because he exercises direct control over the battalion staff. He is also in direct charge of the battalion's main CP in the absence of the commander.

3-68. The battalion staff consists of the operations officer (who also serves as the deputy COS), the assistant operations officer (battalion fire direction officer), the intelligence officer, and the resources officer. The COS also serves as the chief of fire support. The signal section leader also serves as the battalion communications officer. The transport section leader serves as the battalion logistics officer. The staff section operates the main CP. During static operations, staff functions are performed in tents, tunnels, caves, or local accommodations. The command and staff vehicles are then dispersed and camouflaged.

Towed Mortar Battery

3-69. The towed mortar battery normally consists of—
- Headquarters and service section.
- Fire coordination platoon.
- Two towed mortar platoons.
- Supply and transportation section.

3-70. The headquarters and service section consists of a command team, a support team, and a supply and transport team. The battery deputy commander (in the support team) normally is also the battery fire direction officer. The fire direction chief (in the fire control section of the fire coordination platoon) assumes the duties of the fire direction officer in the battery deputy commanders' absence. Depending on mission requirements, either the battery commander or deputy commander remains with the weapons battalion commander to facilitate fire support.

3-71. When required, an observation team (from the observation section of the fire coordination platoon) accompanies a guerrilla unit. In this case, the observation section vehicle drops the team off and returns to the section headquarters. Supported units also provide additional observers.

3-72. The towed mortar battery normally has six 120-mm mortars—two platoons of three mortar sections each. However, some batteries may have six 100-mm mortars. For normal travel, the light truck (prime mover) of each mortar section carries the mortar crew and ammunition for the immediate fire support mission. For movement over short distances, the 100- or 120-mm mortar can be dismantled into three parts: barrel, bipod, and baseplate. Animals or carts can pack the mortar in its three parts. Additional ammunition bearers may be used when required. Civilians or draft animals may also be used as bearers or porters. The guerrillas may have no vehicles at all and depend on caches, porters, draft animals, or other transportation or resupply means.

Note. Another alternative is for the battery to have nine 81- or 82-mm mortars (not towed), by adding a third mortar platoon.

Rocket Launcher Battery

3-73. The rocket launcher battery of the weapons battalion typically consists of—
- Headquarters and service section.
- Fire coordination platoon.
- One multiple rocket launcher (MRL) platoon.
- One rocket launcher platoon.

3-74. The headquarters and service section and the fire coordination platoon have the same structure and functions as in the towed mortar battery (see above). The MRL and rocket launcher platoons may receive additional transportation support from the weapons battalion headquarters or the brigade transport company.

Multiple Rocket Launcher Platoon

3-75. The MRL platoon typically consists of a platoon headquarters and three MRL sections with one 107-mm launcher each. The platoon can be employed as a platoon or broken into separate sections to support guerrilla or HK companies. The 107-mm MRL on a wheeled carriage can be towed by almost any vehicle or pulled by personnel or a yoked animal. Several versions can be broken down into man-packable loads. When moving dismounted for any distance, an MRL section may require augmentation of other guerrillas or civilian personnel to serve as additional ammunition bearers and to move the MRLs.

3-76. Single-tube, man-portable, tripod-mounted launchers are also available. Twelve single-tube 107-mm rocket launchers may replace each 12-tube MRL. When using single-tube rocket launchers, a vehicle is not required. However, extra porters and ammunition bearers are then required. Improvised 107-mm rocket firing pads can be constructed using materials such as dirt, bamboo frames, pipes, or crossed stakes.

Rocket Launcher Platoon

3-77. The rocket launcher platoon typically consists of a platoon headquarters and six rocket launcher sections. Each section operates two 122-mm single-tube rocket launchers. This provides a total of 12 single-tube rocket launchers in the platoon. When moving dismounted for any distance, a rocket launcher section requires augmentation of other guerrillas or civilian personnel to serve as bearers to transport the launchers and additional rockets.

3-78. Typically, each rocket launcher section has one light truck. The truck can carry one or both launchers close to the firing point. The firing team dismounts, transports the launcher to the firing position and prepares it to fire. The team then fires several rockets, probably no more than five, before detection probability is high. Each launcher can fire on individual targets or it can fire preplanned volley fires with other launchers against a single target. The truck (whenever possible) will return and pick up the firing team and launcher. If the truck cannot retrieve the team and launcher, they either join up with another guerrilla unit or exfiltrate to another possible pickup point.

3-79. The 122-mm launcher can be broken into two-one man loads for transport. The 122-mm rocket can also be broken into two-one man loads. Improvised 122-mm. rocket firing pads can be constructed using materials such as dirt, bamboo frames, pipes, or crossed stakes. Caches of rockets may be used for resupply.

Antitank Battery

3-80. The antitank (AT) battery of the weapons battalion typically consists of a headquarters and service section, three antitank platoons, and one antitank hunter-killer (AT HK) platoon. These platoons can be broken down to support guerrilla or HK companies or their subordinates. The headquarters and service section has the same structure and functions as in the towed mortar battery and rocket launcher battery (see above).

Antitank Platoon

3-81. Each AT platoon typically has a platoon headquarters, an antitank guided missile (ATGM) section, and a recoilless gun section. The AT platoon receives transportation support from the battery headquarters and service section or the battalion transport section. Some ATGM and recoilless gun sections may have additional ammunition bearers. Civilians, draft animals, carts, bicycles, and bearers/porters may also be used to assist in transporting the weapons and munitions.

3-82. **ATGM Section.** The ATGM section of the AT platoon typically consists of a section headquarters and four ATGM teams. It may be employed as a section or allocated by team to guerrilla or HK companies. They may also be integrated into AT HK teams at company level. Each ATGM team has one ATGM launcher and carries four ATGMs for its launcher. These and/or additional ATGM launchers may also be dispersed among guerrilla HK teams.

3-83. Each ATGM team of the AT battery's AT platoon typically has four members. This organization is different from an ATGM team in the guerrilla battalion's weapons company, which has three members. The difference is the addition of one grenadier and his ATGL to the team, and the team may be reconfigured based on the available ATGM system.

3-84. **Recoilless Gun Section.** The recoilless gun section of the AT platoon typically consists of a section headquarters and three recoilless gun teams, each with one recoilless gun. It may be employed as a section or allocated by team to guerrilla or HK companies. When towed, a recoilless gun requires a crew of three. When transported solely by the crew, it requires a crew of at least four. Some recoilless gun sections may have additional ammunition bearers.

Antitank Hunter-Killer Platoon

3-85. The AT HK platoon typically consists of a platoon headquarters and three AT HK sections. It may be employed as a platoon or by sections. It may also be broken up into teams and allocated to support separate guerrilla or guerrilla HK units. The platoon's primary weaponry is 18 84-mm recoilless rifles.

3-86. Each AT HK section typically consists of a section headquarters and three AT HK teams. The senior team leader also serves as the assistant section leader. The AT HK section may be employed as a section of six recoilless rifles or allocated by team to guerrilla or HK companies. Teams may also be integrated into AT HK teams at company level.

3-87. Each team of six guerrillas operates two recoilless rifles, which typically are used together for maximum effectiveness. Each AT recoilless rifle gunner carries three rounds (one round in the weapon and two carried on load bearing equipment). The team leader, assistant gunner, and ammunition bearer all carry at least six rounds each and may carry more depending on the type of ammunition. Additional ammunition bearers may be used when required.

Signal Section

3-88. The signal section of the weapons battalion typically consists of a section headquarters, a courier squad, a voice squad, a digital team, and a wire team. The structure and capabilities of this section are likely to be quite similar to those of the signal section in a guerrilla battalion (see below). The signal section leader also serves as the battalion signal officer.

Transport Section

3-89. The transport section of the weapons battalion is likely to be quite similar to one of the transport sections in the transport platoon of a guerrilla battalion. (See below.)

RECONNAISSANCE COMPANY

3-90. The reconnaissance company of a guerrilla brigade typically has a company headquarters and service section, three reconnaissance platoons, and one intelligence and electronic warfare (EW) platoon. This company may be augmented by other guerrilla units, HK teams, or local sympathizers. Members of this company (and augmentees) may be in civilian clothes, in which case they may not have a weapon or radio and would appear to be noncombatants.

Reconnaissance Platoon

3-91. The reconnaissance platoon typically has a platoon headquarters and three reconnaissance squads. A reconnaissance squad can break down into three or four teams. The platoon may be mounted in light trucks or on motorcycles, or may be dismounted based on the tactical situation. Since there are typically not enough vehicles to transport all the teams simultaneously, a squad or team may be dropped off at a point at least part way to the location for its reconnaissance mission and perhaps picked up later at a designated location. Motorcycles can be used to support their individual squad, or they can be grouped together to serve as a high-mobility reconnaissance squad. Squads may be augmented by military or civilian vehicles (motorcycles, trucks, cars, bicycles, carts, or high-mobility all-terrain vehicles) or personnel depending on the mission. Local sympathizers may provide assistance and information.

Intelligence and Electronic Warfare Platoon

3-92. The intelligence and EW platoon typically has a platoon headquarters, a signals reconnaissance section, a global positioning system (GPS) jamming section, an EW squad, and an integration squad. Whenever possible, the intelligence and EW platoon operates out of tents, tunnels, caves, or local accommodations. Its activities are always covered, camouflaged, and concealed.

Signals Reconnaissance Section

3-93. The signals reconnaissance section conducts radio intercept or direction finding (DF) tasks. The section typically consists of a section headquarters, a signal intercept and exploitation squad, and a radio DF squad. Each of the squads consists of an (intercept or DF) base vehicle, a mobile (intercept or DF) position, and three or more dismounted (intercept or DF) teams. The dismounted teams generally accompany guerrilla units and report back to their squad leader.

3-94. The signal intercept and exploitation squad coordinates all intercept and DF operations based on guidance received from the platoon's integration squad, which receives its guidance from the intelligence and EW platoon leader. The platoon leader, in turn, receives his guidance from the reconnaissance company operations officer. The signal intercept and exploitation squad tasks the DF squad by providing target technical data and priority of location. The effectiveness of intercept (HF/VHF/digital) capabilities and/or radio DF depends on how the enemy uses his electronic systems.

Global Positioning System Jamming Section

3-95. The GPS jamming section disrupts enemy GPS receivers. The section typically has two GPS jamming squads. Each squad has a mobile, trailer-mounted GPS jammer. The trailer-mounted jammers are employed separately. Due to their high level of vulnerability, one may be held back for later use. Man-portable GPS jammers are dispersed via two-man teams in civilian clothes and using motorcycles, bicycles, or other civilian vehicles. These teams either transmit while on the move or emplace and operate the jammers. In order to survive when dismounted, however, they must make frequent moves when transmitting (jamming). Another way to increase survivability is to employ the jammers in areas of high civilian density. Additional man-portable jammers can be distributed to other guerrilla units with instructions on how, where, and when to use them. Sappers can infiltrate enemy compounds and surreptitiously emplace man-portable jammers.

Integration Squad

3-96. The integration squad performs signals intelligence (SIGINT) analysis, collection management, and reporting for the intelligence and EW platoon. It receives guidance from the reconnaissance company operations officer through the platoon leader. This squad also disseminates intelligence information based on the analysis of intercepted data and emitter locations.

SAPPER COMPANY

3-97. Sappers are guerrillas trained to perform some functions typically associated with raiders, engineers, or rangers. Sappers are *not engineers*. The sapper company of a guerrilla brigade typically consists of—
- A headquarters and service section.
- Three sapper platoons.
- A transport section.

A guerrilla brigade can form more than one sapper company.

3-98. The structure and capabilities of the sapper platoons are likely to be quite similar to those of the sapper platoon in a guerrilla battalion. (See Sapper Platoon under Guerrilla Battalion, below, for more detail on the nature of sappers and the types of functions they can perform.) The main difference may be in the scope of sapper activities. Sapper platoons, particularly those of a guerrilla brigade, can serve as an independent combat force making deep thrusts from different directions simultaneously. However, a sapper platoon or individual squads may also be allocated to accompany and support a guerrilla battalion, guerrilla company, or HK company. The sappers may also augment other sapper squads, HK teams, and other guerrilla units as necessary.

3-99. Sappers can perform the following functions:
- Infiltrate enemy installations and areas.
- Scout (making accurate diagrams for future attacks).
- Guide other sappers or guerrillas (or affiliated SPF) through enemy lines and obstacles to perform their missions.
- Conduct route reconnaissance.
- Conduct mine warfare.
- Breach obstacles.
- Emplace mines (especially nuisance minefield, IEDs and signal flares).
- Conduct general demolition.
- Emplace field expedient fortifications and obstacles (such as cratering).
- Set side-attack mines.
- Support antitank and countermobility operations.
- Conduct and/or assist in ambushes.
- Employ limited smoke or expedient obscurants.
- Provide general engineer-like support.

See Sapper Platoon under Guerrilla Battalion, below, for more detail on which squad(s) within a sapper platoon typically perform each of these functions. However, all squad members are cross-trained to perform all sapper functions.

3-100. The transport section of a sapper company transports equipment and a basic load of mines and demolition materiel. It includes support personnel armed with light machineguns, who are dispersed among the trucks while moving.

TRANSPORT COMPANY

3-101. The transport company of the guerrilla brigade typically consists of a headquarters and service section and three transport platoons. The structure and capabilities of these platoons are likely to be quite similar to those of the transport platoon of a guerrilla battalion (see below). Some guerrilla brigades may not require a full transport company. Only one or two platoons may suffice.

SIGNAL PLATOON

3-102. The signal platoon of a guerrilla brigade typically consists of a platoon headquarters and two signal sections. Each signal section typically has a section headquarters, a courier squad, voice squad, digital team, and wire team. The structure and capabilities of these sections are likely to be quite similar to those of the signal section of a guerrilla battalion (see below for details). A designated signal platoon truck primarily supports operations conducted in a main CP. It is equipped with a small satellite communications (SATCOM) antenna, digital, and secure communications. Depending on mission requirements, other light trucks and trailers provide the mobility for the platoon and its squads or teams.

MEDICAL PLATOON

3-103. The medical platoon of a guerrilla brigade typically consists of a platoon headquarters and two medical sections. These sections have the same structure and capabilities as the medical section of a guerrilla battalion (see below). The platoon provides immediate medical care, trauma stabilization, and minor surgical actions. An officer (physician) or medical assistant (physician's assistant) and senior noncommissioned officers provide limited medical intervention, minor surgery, and treatment. Guerrilla commanders may mobilize local medical personnel from the population to treat their ill and wounded. The guerrilla force has a limited inpatient capability. Whenever possible, medical functions are performed in tents, tunnels, caves, or local accommodations. Severe and longer-term care relies on evacuation to civilian, military, or other medical facilities. More routine and excess ill and wounded are backhauled in general-purpose cargo vehicles.

3-104. Guerrilla medical support is coupled with local medical assets in the area when available. Maximum use is made of local medical assistance and facilities regardless of capability. In some instances, the brigade medical platoon may attempt to operate in conjunction with the assets of a village clinic. Local sympathizers may volunteer their homes, equipment, vehicles, and services. They may also assist in the evacuation of wounded guerrillas to civilian facilities.

3-105. Medical vehicles may be a mix of military and civilian, or all civilian. Carts may also be used to transport wounded. Cargo trailers transport medical equipment and supplies, and in emergencies may transport wounded. Supported guerrilla units receive liters from the medical section to transport wounded. The guerrilla unit provides its own liter bearers.

3-106. Guerrilla medics are *combatants*, trained to fight as guerrillas. They fight when necessary to support the guerrilla mission. A medical aid station is usually set up at battalion level while other medics accompany guerrillas in the fight. Medical platoon personnel may be a mixture of men and women. Women may comprise at least 50 percent or more of the medical platoon strength.

GUERRILLA BATTALION

3-107. The composition of a typical guerrilla battalion is as follows:
- Battalion headquarters.
- Guerrilla companies or guerrilla HK companies.
- Weapons company.
- Reconnaissance platoon.
- Sapper platoon.
- Transport platoon.
- Signal section.
- Medical section.

Figure 3-3 shows an example of a guerrilla battalion with two task-organized HK companies.

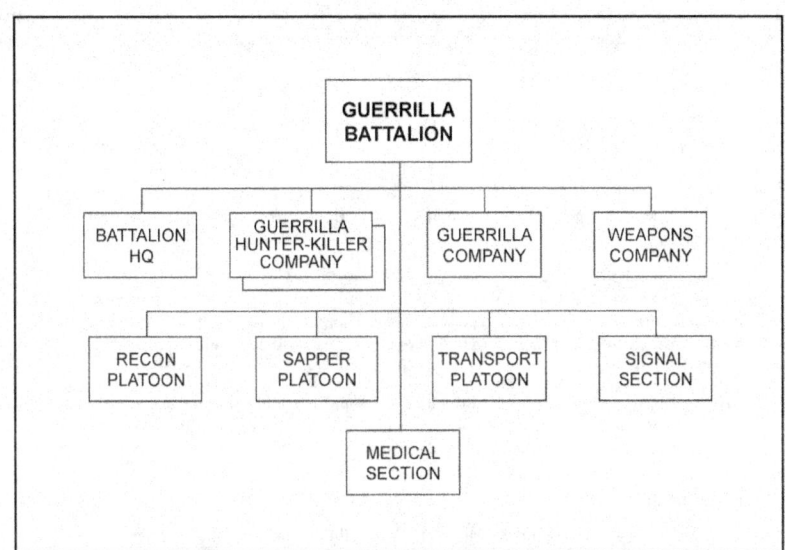

Figure 3-3. Guerrilla battalion (example)

Note. HK company organizations configured from subordinate units of a guerrilla battalion are discussed later in this chapter.

3-108. A guerrilla battalion may have as many as 1,000 guerrillas, but could be considerably smaller. Each guerrilla battalion differs in capability, but all battalions have a similar structure of a battalion headquarters and combat, CS, and CSS units. Depending on the number of guerrilla and/or HK companies, the size and strengths of support units can also vary.

3-109. Often a battalion-size guerrilla unit will be tailored for a particular OE or mission. If task-organized, capabilities may include four or five guerrilla or guerrilla HK companies, some organic battalion units, a task-organized, composite weapons battery from a guerrilla brigade, and possibly intelligence and EW support. A composite weapons battery could have a mix of mortar, rocket launcher, and antitank platoons. Guerrillas can also coordinate and act in concert with local insurgent and/or higher insurgent organizations.

3-110. A guerrilla battalion may be any combination of guerrilla companies or guerrilla HK companies. When a battalion consists predominantly of guerrilla HK companies, its designation is a guerrilla HK battalion. The HK company is especially effective in close environments such as urban, forest, or swamp terrain that offers opportunities to canalize the enemy into preplanned kill zones. The HK company fights as HK groups, sections, and teams. The guerrilla company fights as platoons, squads, and fire teams. With sufficient equipment and resources, it can conduct sustained company-level actions.

GUERRILLA BATTALION HEADQUARTERS

3-111. The guerrilla battalion headquarters consists of a command section and a staff section. During static operations, command and staff functions may be performed in tents, tunnels, caves, or temporary and/or permanent local accommodations.

Command Section

3-112. The command section comprises the battalion commander, deputy commander, and the immediate support personnel. The deputy commander's light truck may be used as a CP.

Staff Section, Guerrilla

3-113. The battalion chief of staff (COS) is part of the command group and staff, but is normally located with the staff section in order to exercise direct control over the staff. He also controls the main CP in the absence of the commander. The battalion staff consists of the operations officer (who also serves as the deputy COS), the assistant operations officer, the intelligence officer, and the resources officer. Each of these primary staff members heads a small staff section consisting of himself and one or two other staff personnel.

3-114. The commander of the weapons company also serves as the chief of fire support. The signal section leader also serves as the battalion communications officer. The platoon leader of the reconnaissance platoon serves as the battalion chief of reconnaissance. The platoon leader of the transport platoon serves as the battalion logistics officer. When the guerrilla battalion is affiliated with or subordinate to an insurgent organization, the staff section may also have a political advisor.

3-115. Any liaison teams attached to the battalion would be under the operations officer. Liaison may include SPF personnel for particular missions or training. Liaison may also be present from regular force units operating in coordination with the guerrilla force.

WEAPONS COMPANY

3-116. The weapons company of a typical guerrilla battalion consists of the following:
- Headquarters and service section.
- Fire coordination platoon.
- Mortar platoon.
- Multiple rocket launcher (MRL) platoon.
- Antitank platoon.
- Man-portable air defense system (MANPADS) squad.

Headquarters and Service Section

3-117. The headquarters and service section of the weapons company is very similar to that of the guerrilla company, but with some additional transport and communications equipment. It typically consists of a command team, a support team, and a supply and transport team. The weapons company commander locates where he can best assist the battalion commander, which may often mean being colocated with the battalion commander. The deputy commander of the weapons company (in the support team) normally is the fire direction officer.

Fire Control Coordination Platoon

3-118. The fire coordination platoon typically consists of the platoon headquarters, a fire control section, and an observation section. The fire control section serves as the company fire direction center. The fire direction chief (in the fire control section) assumes the duties of the fire direction officer in the absence of the company deputy commander. The fire control section may have a laptop-based automated fire control system (AFCS). When available, the AFCS provides fast and accurate ballistic computations of artillery and rocket launcher weapons, fire support coordination, message transfer in digital format and ammunition accounting at the weapons company level. When required, an observation team accompanies a guerrilla unit. Supported units also provide additional observers.

Mortar Platoon

3-119. The mortar platoon typically consists of a platoon headquarters and three mortar sections. The sections normally consist of a section headquarters and two mortar squads, each with one 82-mm (or 81-

mm) mortar. A mortar squad normally consists of a squad leader, a gunner, an assistant gunner, and four ammunition bearers. When moving dismounted, the section may require augmentation to serve as additional ammunition bearers. The mortar section may be augmented with high-mobility all-terrain vehicles, depending on the mission. 82-mm mortars can fire 82-mm and 81-mm ammunition including extended-range munitions. However, this platoon may substitute 100- or 120-mm mortars for the 81- or 82-mm mortars. As with its 81- and 82-mm counterparts, the 100-mm mortar, when broken down in three parts, can be man-packed or animal-carried, or the complete weapon can be carried on a light vehicle. The 120-mm mortars may also be substituted; however, this heavy mortar requires a tactical vehicle or light truck to transport each mortar.

3-120. Some mortar platoons may have only four mortars, consisting of two mortar sections of two mortars each. The mortar platoon receives transportation support from the either the headquarters and service company or the battalion transport platoon. Additional ammunition bearers may be used when required.

Multiple Rocket Launcher Platoon

3-121. The MRL platoon typically consists of a platoon headquarters and three MRL sections, with one MRL per section. It can be employed as a platoon or broken down into separate sections to support subordinate companies or HK groups. The platoon receives transportation support from the either the battery's headquarters and service company or the battalion transport platoon. The wheeled MRL carriage can be towed by almost any vehicle or pulled by personnel or a yoked animal. In some cases, the MRL can be mounted on a vehicle. When moving dismounted for any distance, an MRL section may require augmentation of personnel to serve as additional ammunition bearers or to move the MRL. Some versions of the 107-mm MRL can be broken down into man-packable loads. Light trucks that serve as prime movers for the towed MRLs may also transport other assets of the weapons company, such as mortars or recoilless guns.

3-122. In lieu of MRLs, some platoons may have single-tube rocket launchers. In that case, 12 single-tube launchers replace each 12-tube MRL, and extra porters or ammunition bearers are required. Improvised 107-mm rocket firing pads can be constructed using dirt, bamboo frames, or crossed stakes.

Antitank Platoon

3-123. The AT platoon typically consists of a platoon headquarters, an antitank guided missile (ATGM) section, and a recoilless gun section. The AT platoon normally has no vehicles. The ATGM section and recoilless gun section may receive transportation support from the weapons company's headquarters and service section or the battalion transport platoon. They may also be transported on the vehicles belonging to the MRL platoon.

3-124. The ATGM section typically consists of a section headquarters and three ATGM teams. It may be employed as a section or allocated by team to guerrilla or HK companies. These teams may also be integrated into HK teams throughout the battalion. One person can fire the man-portable ATGM; however, the crew normally consists of an ATGM gunner (team leader) and assistant gunner. The team may also include one or more riflemen to provide security and/or carry additional missiles.

3-125. The recoilless gun section typically consists of a section headquarters and three recoilless gun teams. It may be employed as a section or allocated by team to guerrilla or HK companies. A team normally consists of a gunner (team leader), an assistant gunner, and a varying number of ammunition bearers. A 73-mm recoilless gun is man-portable, but is usually carried in a vehicle or cart. Some versions may have removable wheels; when towed, it requires a crew of three. When transported only by the crew, the weapon requires a crew of at least four, including an additional ammunition bearer. Some recoilless gun teams may use civilian personnel and vehicles from the local population to assist in transporting the weapon and ammunition.

MANPADS Squad

3-126. The MANPADS squad typically has two MANPADS launchers and one heavy machinegun. The entire MANPADS squad may be employed together, or it can be broken down into three teams. In the latter option, two teams with a single MANPADS launcher each and one team with the heavy machinegun can be deployed with guerrilla or HK companies.

3-127. If the squad has a light truck, the vehicle can carry up to five missiles for each MANPADS launcher. When dismounted, each MANPADS gunner carries a gripstock launcher and one missile. The MANPADS assistant gunner carries one additional missile; he also carries the electronic plotting board and enters location and direction data of approaching targets. Local civilian personnel and vehicles may transport additional missiles. The senior MANPADS gunner is the assistant squad leader.

RECONNAISSANCE PLATOON

3-128. The reconnaissance platoon of a guerrilla battalion typically consists of a platoon headquarters and three reconnaissance squads. Each squad can break down into two or three teams. One scout per squad is designated as the sniper (or marksman) and uses a sniper rifle.

3-129. Based on the situation, the platoon may be dismounted or mounted in light trucks or on motorcycles. Surveillance and communications equipment is man-portable, but some equipment may require vehicle support. Some scouts may be assigned additional duty as a motorcycle or light truck drivers. The reconnaissance squads may not be able to lift all their personnel and equipment at one time without augmentation. The squads may be carried part way on a patrol and dismounted. They are then picked up on the way back. Depending on the mission, the platoon may also be augmented with military or civilian vehicles (trucks, cars, motorcycles, bicycles, carts, or high-mobility all-terrain vehicles) or with additional personnel and equipment from guerrilla companies.

3-130. Each reconnaissance squad normally has a motorcycle, which extends the range of the squad. Motorcycles can be used to support their individual squad, or they can be grouped together to serve as a high-mobility reconnaissance squad. For longer patrols, the motorcycles can be loaded into a light truck.

3-131. Each reconnaissance squad is equipped with long-range cordless telephones (LRCTs). The LRCT base station remains with the squad vehicle and serves as a retransmission station for the patrols. Digital cameras, camcorders, or computers may be linked to the LRCTs for transmittal to the platoon headquarters or battalion via the squad base station. The LRCT base station vehicle must remain within range of the LRCTs used by the squad or by teams within the squad.

3-132. Reconnaissance personnel may wear civilian clothes and use common civilian model trucks, cars, motorcycles, mopeds, motor scooters, or bicycles. In this case, the guerrilla would probably not have an overt weapon or radio and would appear to be a noncombatant. Local sympathizers may provide assistance and information.

SAPPER PLATOON

3-133. Sappers are guerrillas trained to perform some functions typically associated with raiders, engineers, or rangers. Sappers are *not engineers*. Guerrilla battalions may use sappers in an assault and/or demolition role. In a raider role, sappers are the lead or primary (assault) element in an assault on fixed installations or military field positions. Armed primarily with explosives charges, sappers breach the defensive perimeter and neutralize designated positions in advance of the attacking main body. The sapper unit can serve as an independent combat force making deep thrusts from different directions in the enemy-held region. At other times, they can accompany and support guerrilla and HK missions throughout an AOR. The sappers also augment other sapper squads, HK teams, and other guerrilla units as necessary. Sappers may also serve as a stay-behind or independent unit to conduct disruption operations. The sapper platoon also coordinates suicide bombings.

3-134. Guerrilla sappers (or their SPF advisors from a regular military force) can also train local civilians to be sappers. The sappers may be a mix of men, women, and children. Women and children may be used as runners, messengers, scouts, guides, drivers, porters, fighters, suicide bombers, lookouts, or in several other roles.

3-135. A guerrilla battalion may form more than one sapper platoon. A sapper platoon typically is organized as—

- A platoon headquarters.
- An infiltration and scout squad.
- A mine warfare and demolition squad.
- An improvised explosives and signal flares squad.
- A general support squad.

Text below outlines the functions normally associated with each of these squads. However, all squad members are trained to perform all sapper functions. Each squad normally carries a mix of mines and demolitions. The mix varies according to the mission. A trailer may transport equipment and a basic load of mines and demolitions and may be dropped off to be recovered later.

Infiltration and Scout Squad

3-136. The infiltration and scout squad serves as a sapper/raider element and may—
- Infiltrate enemy installations and areas.
- Serve as the lead or primary (assault) element in an assault on a fixed installation or a military field position.
- Set demolitions and side-attack mines.
- Serve as scouts (making accurate diagrams for future attacks).
- Conduct route reconnaissance.
- Guide other sappers or guerrillas (or affiliated SPF) through enemy lines and obstacles to perform their missions.
- Conduct and/or assist in ambushes.

A truck may transport squad personnel to an appropriate dismount point and return to the platoon.

Mine Warfare and Demolition Squad

3-137. The mine warfare and demolition squad may—
- Conduct mine warfare.
- Breach obstacles.
- Emplace mines (especially nuisance minefield, IEDs and signal flares).
- Support antitank and countermobility operations.
- Scatter mines using a man-portable mine-scattering system.
- Lay controlled minefields.
- Serve as the lead or primary (assault) element in an assault on a fixed installation or a military field position.
- Infiltrate enemy installations and areas.

A trailer typically transports equipment and a basic load of mines and demolitions. It may be dropped to be recovered later. Minelaying operations may require a truck on-site to off-load the mines.

Improvised Explosives and Signal flares Squad

3-138. The improvised explosives and signal flares squad may—
- Manufacture improvised explosive devices (IEDs) and signal flares. (Signal flares are actually a subcategory of IEDs.)
- Emplace and/or detonate the IEDs. (Signal flares are triggered by the unsuspecting.)
- Provide IEDs (or signal flares) to other trained sappers or guerrillas to emplace and/or detonate.
- Serve as the lead or primary (assault) element in an assault on a fixed installation or a military field position.
- Infiltrate enemy installations and areas.
- Augment other sapper squads and HK teams as necessary.
- Coordinate suicide bombings.

Chapter 3

> *Note.* Guerrillas (and other irregular OPFOR) commonly use IEDs as secondary devices to detonate on the arrival of responding personnel. IEDs can be detonated by a variety of means, including remote, command, electrical trip wire, pressure, time, and others.

General Support Squad

3-139. The general support squad may—
- Provide general engineer-like support to the sapper platoon or the guerrilla battalion.
- Employ limited smoke or expedient obscurants.
- Provide water purification and minor construction.
- Emplace field expedient fortifications and obstacles (such as cratering).
- Conduct general demolition.
- Conduct assault breaching.
- Emplace anti-vehicle wire obstacles.
- Augment other sapper squads and HK teams as necessary.
- Serve as the lead or primary (assault) element in an assault on a fixed installation or a military field position.
- Infiltrate enemy installations and areas.

TRANSPORT PLATOON

3-140. The transport platoon of a guerrilla battalion typically has a platoon headquarters and two transport sections. The platoon leader serves as the battalion logistics officer.

3-141. Each transport section operates with three light trucks and five medium trucks. Single- and two-axle trailers provide additional cargo capacity. Water and POL trailers are augmented with flexible storage bladders for both water and fuel. The transport platoon assists in transporting assets of the battalion's weapons company (mortars, rocket launchers, antitank weapons, and munitions) whenever necessary. Whenever possible, corner-mounted mechanical hoists are used to load and unload the vehicles.

3-142. Vehicles are generally a mix of older civilianized military and civilian models, or all civilian models. They will not appear to be or look like military vehicles and are intended to be indistinguishable from civilian vehicles. The vehicles are kept as dispersed as possible, in order to prevent detection and destruction by enemy forces. Rarely, if ever, will all vehicles in the transport platoon be colocated. Vehicles may be dispersed for use by locals as commercial, delivery, agricultural, general cargo, construction, militia, or general-purpose vehicles in everyday life. When required, the transport platoon leader (battalion logistics officer) or the battalion resources officer will assemble the appropriate mix of vehicles to transport specific items to a specific location. The vehicles can then melt back into the civilian population when not assembled to support guerrilla operations. Guerrillas do not use static truck parks or assembly areas for vehicles.

3-143. The cargo trucks normally carry materiel to meet only the immediate combat needs of the guerrilla battalion. At other times, they may transport materiel for caches and prepositioning. When necessary, additional transportation support may be received from higher headquarters or external sources. Transportation can be augmented with vehicles, draft animals, or bearers/porters requisitioned or confiscated from the local citizenry. Local sympathizers may volunteer their equipment and services. Local civilian mechanics service and/or repair the vehicles when required.

SIGNAL SECTION

3-144. The signal section of a guerrilla battalion typically consists of a section headquarters, a courier squad, a voice squad, a digital team, and a wire team. The signal section leader also serves as the battalion communications officer.

3-145. The courier squad typically uses motorcycles, mopeds, motor scooters, or bicycles of the same types commonly used by the local population. High-mobility or all terrain vehicles may be substituted for motorcycles. The courier may be a male or female in civilian clothes. In this case, the courier probably will not have an overt weapon or radio and will appear to be a noncombatant. Another guerrilla may ride with the courier as a lookout or to provide security. (In the latter case, the additional rider could have a rifle or light machinegun.) Depending on the circumstances, some couriers may not use vehicles at all. Message transmission can be written, digital, or memorized and presented orally to a designated recipient. The courier squad can also conduct drive-by shootings or attacks.

3-146. The voice squad supports operations conducted in the battalion headquarters. A signal truck primarily supports the battalion main CP. It is equipped with a small SATCOM antenna and can provide digital and secure communications. Depending on mission requirements, organization, and terrain, two light trucks may be substituted for the signal truck. Some staff communications are remoted back to the signal section.

3-147. The digital team is responsible for all digital communications between the battalion command section and staff, to other battalion subordinates, and to higher headquarters. It also provides a remote communications capability. The team is equipped with a small SATCOM antenna and long-range cordless telephones, and provides secure communications.

3-148. The wire team is responsible for all wire (landline) communications between the battalion command section and staff and the subordinate units. It also provides a remote communications capability. Landline communications are used whenever possible.

MEDICAL SECTION

3-149. The medical section of a guerrilla battalion is responsible for temporary medical treatment. A medical aid station is usually established at battalion level with a number of medics directly supporting guerrillas in the fight. Supported guerrilla units receive liters from the medical section to transport wounded. The guerrilla unit provides its own liter bearers. Cargo trailers transport medical equipment and supplies. In emergencies, the trailers may transport wounded. These trailers may be dropped at the aid station when light trucks serve as ambulances. Evacuation and transportation may be a mix of military and civilian vehicles, wagons, or carts. Whenever possible, medical functions are performed in tents, tunnels, caves, or local accommodations. Local medical support may be available. Civilian sympathizers may volunteer use of their facilities, vehicles, and services, or may be coerced into providing temporary medical assistance. (For additional information on medical support, see the Medical Platoon under Guerrilla Brigade, above.)

Note. Guerrilla medics are *combatants*, trained to fight as guerrillas. They fight when necessary to support the guerrilla mission.

GUERRILLA COMPANY

3-150. A typical guerrilla company consists of—
- Headquarters and service section.
- Three guerrilla platoons.
- Weapons platoon.

HEADQUARTERS AND SERVICE SECTION

3-151. The headquarters and service section of the company has a command team, support team, and supply and transport team. The command team consists of the company commander, a radio telephone operator (RTO) and a messenger-runner. The support team is the deputy commander, first sergeant, an RTO, and a medic. The supply and transport team consists of the supply sergeant, drivers, and trucks that provide the company a degree of autonomy from the guerrilla battalion.

3-152. Figure 3-4 shows an example of such a company. A guerrilla company fights as platoons, squads, and fire teams (see Guerrilla Company, below).

Chapter 3

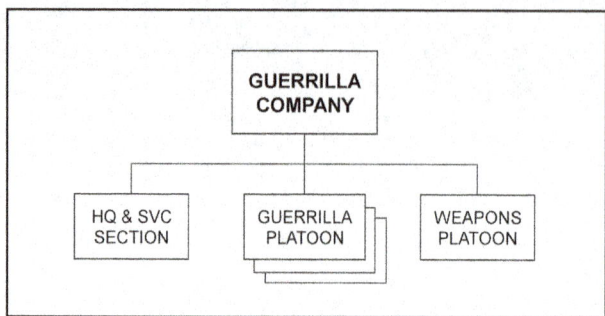

Figure 3-4. Guerrilla company (example)

WEAPONS PLATOON

3-153. The weapons platoon of a guerrilla company typically consists of a platoon headquarters, a mortar section, a recoilless gun section, a machinegun section, and a sniper section. Weapons of this platoon provide a heavy volume of fires and extend the range of weapons effects beyond the range of guerrilla platoon weapons. The platoon receives additional transportation support from the supply and transport team of the company's headquarters and service section. Unless vehicles are available, ammunition for the platoon's mortars and recoilless guns may require additional ammunition bearers and may be distributed among members of the company. This platoon may also allocate weapons to individual guerrilla platoons. When a guerrilla company is task-organized into an HK company, the weapons platoon typically ceases to exist as a separate unit. Its weapons are then redistributed among various parts of the HK company.

Mortar Section

3-154. The mortar section typically consists of three mortar squads. The mortar section leader serves as the squad leader of one of the three motor squads. The mortar section may be employed as a section or allocated by squad to guerrilla platoons.

Recoilless Gun Section

3-155. The recoilless gun section typically consists of a section headquarters and three recoilless gun teams. It may be employed as a section or allocated by team to guerrilla platoons.

Machinegun Section

3-156. The machinegun section typically consists of a section headquarters and three machinegun teams. It may be employed as a section, or it may be allocated to support separate guerrilla platoons, in addition to their own machinegun sections. Depending on the tactical circumstances, some machinegun sections may have additional ammunition bearers.

Sniper Section

3-157. The sniper section typically consists of two sniper teams. The section leader serves as the team leader of one sniper team. Each sniper team consists of a team leader/observer (spotter), a sniper (shooter)/ target designator, and an assistant sniper. The assistant sniper provides additional security, transports equipment, and may serve as a backup for the sniper or observer. See chapter 16 of TC 7-100.2 for the differences between snipers and marksmen and between snipers (or marksmen) in guerrilla forces and those found in regular military forces.

3-158. The primary mission of the sniper section is to serve in a sniper-countersniper role. Depending on mission requirements, a sniper team may also serve as a reconnaissance element, stay-behind element, or part of an HK team.

Note. When a guerrilla company is restructured into an HK company, the whole sniper section typically becomes part of the HK company's headquarters and command section (a section within a section). However, the section may allocate some or all of its snipers to be part of HK teams.

GUERRILLA PLATOON

3-159. A guerrilla platoon typically consists of a platoon headquarters, three guerrilla squads, and a machinegun section. Figure 3-5 shows an example of such a platoon. The size of the guerrilla platoon can vary, and its weapons, equipment, and manning can be tailored for specific missions. One guerrilla in each platoon is typically cross-trained as a medic in addition to primary duty as a rifleman.

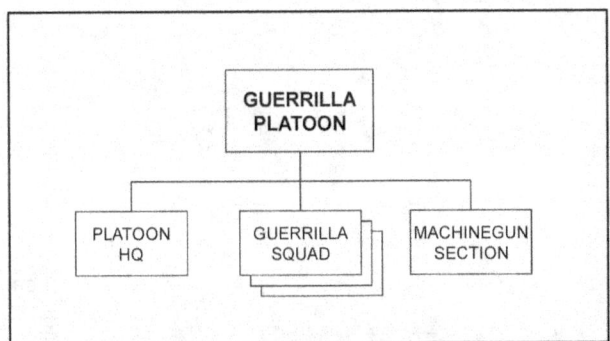

Figure 3-5. Guerrilla platoon (example)

GUERRILLA SQUAD

3-160. The guerrilla squad consists of at least two maneuver fire teams (or HK teams). The squad leader normally commands one fire team while the assistant squad leader commands a second fire team. Each team typically consists of the leader, a machinegunner, a grenadier, an assistant grenadier, and either two rifleman or one rifleman and a guerrilla designated as a sniper or marksman. (See chapter 16 of TC 7-100.2 for the differences between snipers and marksmen and between snipers [or marksmen] in guerrilla forces and those found in regular military forces.) The squad may be augmented by elements from the machinegun section to create three or possibly four maneuver fire teams or HK teams.

3-161. Fire teams structure around one machinegun with remaining team members normally equipped with assault rifles. The riflemen support squad members with other weapons, including the machinegunner, the grenadier, a sniper or marksman, or augmentation from the guerrilla platoon's machinegun section or the company's weapons platoon.

MACHINEGUN SECTION

3-162. The machinegun section, consisting of three teams. It may be employed as a section, or it may be allocated to support separate guerrilla squads. This tailoring of capability enables a guerrilla squad to operate with three or possibly four maneuver fire teams or HK teams. When a guerrilla platoon is restructured as an HK group, the personnel and equipment of the machinegun section typically are dispersed among HK sections and teams.

GUERRILLA HUNTER-KILLER COMPANY

3-163. The guerrilla company can be augmented and restructured into a guerrilla hunter-killer (HK) company made up of numerous small HK teams. Those teams are typically organized into HK sections and the sections into HK groups. The HK team structure is ideal for dispersed combat such as fighting in urban areas and can provide similar capabilities in rural terrain when cover and concealment and channelized avenues favor the guerrilla. Tailored HK units are usually a company-level configuration; however, complete battalions and brigades can be organized for combat as HK units.

3-164. An HK company is based on the personnel and equipment originally found in a guerrilla company. However, the HK company may have additional equipment due to the dispersed nature of HK team employment. For example, it typically would have additional antitank disposable launchers and flame weapons. It may also have three additional 60-mm mortars, possibly dispersed to one team in each HK group. These additional weapons do not necessarily require additional personnel.

3-165. The guerrilla company task-organized as an HK company typically consists of a headquarters and command section and three HK groups. Typically, each HK group has four HK sections, and each HK section has three HK teams. Figure 3-6 shows an example of such a company.

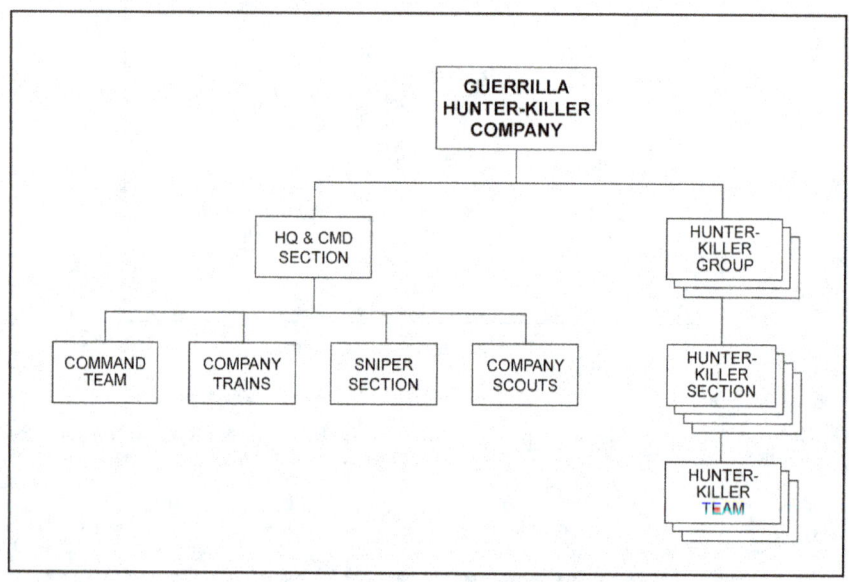

Figure 3-6. Guerrilla hunter-killer company (example)

3-166. An HK company structured as in figure 3-6 can contain a total of 36 HK teams. If the two sniper teams and the company scouts in the headquarters and command section are counted, the HK company can have a total of 39 HK teams.

Note. When a guerrilla platoon is task-organized into an HK group, its machinegun section ceases to exist as a separate unit. Its personnel and equipment are distributed among HK sections and teams. Likewise, when a guerrilla company is restructured into an HK company, the weapons platoon typically ceases to exist as a separate unit. Its weapons are then redistributed among various parts of the HK company.

HEADQUARTERS AND COMMAND SECTION

3-167. The headquarters and command section of an HK company typically comprises the command team, company trains, a sniper section, and company scouts. The company commander and the rest of the command team locate with HK fighting elements. The deputy commander remains in charge of the company trains. The former weapons platoon sergeant now serves as the senior sergeant in the trains. The former weapons platoon leader now performs as a "deputy commander for tactics," as part of the command team. When required, the sniper section and company scouts provide flexible capabilities for additional HK teams.

Sniper Section

3-168. The primary mission of the sniper section is to serve in a sniper-countersniper role. The sniper section consists of two sniper teams. Each sniper team has three members. The section leader serves as the team leader of one sniper team, and each team leader also acts as the observer/spotter. Other team members are a sniper (shooter)/target designator and an assistant sniper. The assistant sniper provides additional security, transports equipment, and may serve as backup for other team members. A sniper team may also serve as a reconnaissance element, a stay-behind element, or as part of an HK team.

> *Note.* When a guerrilla company is restructured into an HK company, the sniper section of the former weapons platoon typically becomes part of the HK company's headquarters and command section (therefore a section within a section).

Company Scouts

3-169. The company scouts typically comprise one team of four members, three of whom normally come from the former weapons platoon of a guerrilla company. The scout team leader and senior scout is the former recoilless gun section leader. The assistant team leader is the former machinegun section leader. The scout who acts as radio telephone operator (RTO) is the former RTO for the weapons platoon. The remaining scout is a former supply specialist with the headquarters and services section of the guerrilla company.

HUNTER-KILLER GROUP

3-170. An HK group is basically a task-organized guerrilla platoon. The HK group typically has a group headquarters and four HK sections, each with three HK teams. (See figure 3-6 on page 3-26.) HK teams in each HK section may vary in their manning, weapons, and equipment. Equipment may be transferred among HK sections and teams. As in the guerrilla platoon, one of the guerrillas in the group is cross-trained as a medic.

SECTIONS ONE, TWO, AND THREE

3-171. Sections one, two, and three are typically formed by augmenting the three squads of a guerrilla platoon with personnel and equipment from that platoon's machinegun section. Each of these three sections has three teams, designated as HK teams one, two, and three. Each of the subordinate teams may differ, as in the following examples:
- The section leader also serves as team leader of team one.
- Team one has a designated sniper or marksman.
- Teams one and two each have a grenadier and assistant grenadier.
- All three teams have a machinegunner, and team three, based on a machinegun team from the machinegun section, also has an assistant machinegunner for its crew-served weapon.
- Teams two and three have other riflemen who also have duties as ammunition bearers or operators of antitank disposable launchers or flame weapons.

SECTION FOUR

3-172. Section four has a different task-organized HK structure because it comes from the guerrilla company's weapons platoon. The weapons platoon had three machineguns, three mortars, and three recoilless guns. One of each of these weapons goes to the section four of each of the three HK groups in the HK company. In section four, these weapons form the basis for team one (machinegun), team two (mortar), and team three (recoilless gun). The teams may receive additional transportation support from the company trains in the headquarters and command section of the HK company.

TACTICS AND TECHNIQUES

3-173. Functional tactics (see chapter 7) are characteristic of a guerrilla force. However, guerrillas may also use terrorism tactics and techniques (see chapter 6). Functional tactics and terrorism are not mutually exclusive. Guerrillas can use both forms of violent action simultaneously. Guerrillas also apply a full range of information warfare (INFOWAR) capabilities to exhaust enemy resolve (see appendix A).

FLEXIBILITY

3-174. Depending on conditions, guerrilla actions can include a wide spectrum of offensive and defensive actions. Guerrilla organizations are capable of independent actions or may be affiliated with regular military forces and/or insurgent organizations. Actions may be conducted in close coordination with regular military forces when such support is available. In a long-term conflict, this close association with regular military forces can lead to integration of highly trained guerrilla forces into the regular military forces. Specific weapons, equipment, manning, and materiel provide the level of combat power or sophistication required for a particular situation. Guerrilla actions are characterized by elusiveness, surprise, and brief, violent action. Guerrillas can use a broad range of tactics, from terrorism and sabotage through functional tactics similar to those used by regular military forces. This enables them to escalate or deescalate activity almost at will. They apply whatever tactics and techniques best fit the specific situation.

3-175. Guerrilla unit actions may be as large as several brigade-level operations or be as small as a platoon raid or independent HK team ambush. Guerrilla organizations and capabilities can be tailored for brigade, battalion, company, platoon, and squad or team levels of action. The objective of guerrilla actions by small independent forces is often to harass, delay, or disrupt enemy operations. The objective of small forces may be to inflict casualties and damage on the enemy rather than to seize and defend terrain. In the latter case, operations are characterized by the extensive use of surprise.

FUNCTIONAL TACTICS

3-176. Guerrillas, as part of the irregular OPFOR, use variants of the same functional tactics described in TC 7-100.2 for smaller regular military units or SPF. (See chapter 7 for more detail on these functional tactics and examples of how guerrillas can use them.)

3-177. When guerrilla forces first become operational, they usually engage in limited or small-scale activities and operations. If they reach more sophisticated levels of organization, equipment, and training, then they may conduct larger operations using more complex forms of functional tactics, perhaps in conjunction with regular military forces.

Offensive Action

3-178. Guerrillas typically use hit-and-run attacks by lightly armed, small forces. Their tactics emphasize ambushes, raids, snipers, rocket and mortar attacks, and the use of explosive devices. Guerrilla actions are generally offensive, not defensive, and are often harassing in nature. Guerrillas seldom attempt to seize and defend physical objectives and, in general, avoid decisive engagement, unless they know they can win. Their overall aim is often to cause confusion, to destroy infrastructure or security forces, and to lower enemy morale. Guerrilla harassment attempts to keep enemy forces on the defensive and weaken them, which can include destroying resources and disrupting lines of communication (LOCs). One advantage of harassment is that it may create the perception that the guerrillas can strike anywhere and that the enemy cannot prevent it. In rural areas, guerrillas may seize a remote area or conduct raids and small-scale attacks on remote targets and LOCs.

3-179. Even when small guerrilla units are under the command of a guerrilla battalion or brigade, this does not necessarily mean a mass concentration of personnel. On the contrary, small units remain dispersed throughout the area. When enemy forces outnumber the guerrillas, the guerrillas seek to attain local numerical superiority. If guerrillas can successfully concentrate, they can attain victory over small elements of enemy forces. Guerrillas often use simple techniques of speed, surprise, maneuver, and especially infiltration. (See Infiltration, below.)

3-180. Guerrillas use dispersion during their movements. However, near the target area, small guerrilla units mass and then conduct offensive actions. (See Swarming, below.) While the guerrillas are outnumbered by enemy forces, they seek to attain local numerical superiority. In this way, they can attain victory over small elements of enemy forces. The most common techniques employed by guerrillas are the ambush, raid, and small-scale assaults or reconnaissance attacks. These techniques usually target security posts, small forces, facilities, and LOCs. These tactics, if successful, may compel enemy forces to commit larger elements to defensive tasks. Guerrillas can employ some of the types of offensive action also used by smaller tactical units of the regular OPFOR. (See chapter 7 for basic discussion of these offensive actions.) Such actions can include—

- Ambush.
- Assault.
- Raid.
- Reconnaissance attack.

Defensive Action

3-181. Guerrilla strength, equipment, and training is almost always inferior to that of the enemy. Therefore, unless forced to do otherwise by the enemy, guerrilla forces hold defensive positions only for brief periods in support of other actions by guerrilla or affiliated regular forces. Guerrilla forces usually defend themselves by flight or dispersion, by withdrawals, or by creating diversions. Whenever possible, defensive actions are accomplished by offensive raids and ambushes against the enemy's flanks and rear. Guerrillas may also try to ambush the initial enemy attack force to inflict maximum casualties.

3-182. When faced with a large-scale enemy offensive action, a guerrilla commander may chose to—

- Defend the area against attack.
- Disperse units or individuals until the enemy offensive is over.
- Conduct diversionary activities in other areas.
- Withdraw into another area not likely to be included in the enemy offensive.
- Withdraw in a wide, circling movement and then attack against the enemy's rear and base installations.

3-183. The object of the defense may be to make the attack so expensive for the enemy that he will soon abandon it and will not wish to try it again. The guerrilla force seeks to avoid being pinned down so that it can be encircled and destroyed. As the enemy overcomes various defensive positions, the defenders withdraw to subsequent defensive positions or break up and infiltrate through the enemy's lines and attack his rear, flanks, and supply installations. The principles of guerrilla defense of fixed positions are the same as those applicable to regular forces, except that there are fewer supporting fires and that counterattacks are not practicable. Maximum use is made of complex terrain, defensive works, and mines.

Guerrillas can employ some of the types of defensive action also used by smaller tactical units of the regular OPFOR. (See chapter 7 for basic discussion of these defensive actions.) Such actions can include defense of a simple battle position and/or defense of a complex battle position.

TERRORISM

3-184. Guerrillas can also use terrorism achieve psychological impact. See chapter 6 for terrorism tactics and techniques that guerrillas can use.

INFILTRATION

3-185. Land infiltration involves the use of various modes of transportation or techniques such as commercial vehicles, railway trains, or infiltration on foot, possibly along with refugees. Before the mission, the guerrilla unit is briefed on the known locations of selected individuals who will furnish assistance and on the established means of contacting them. These individuals may be used as local guides and sources of information, food, and shelter. Since there are local sources for survival items, the unit can restrict the equipment and supplies to be carried to mission-essential items (individual arms, equipment, and communications gear).

3-186. A very successful infiltration method used by guerrillas (and/or affiliated SPF) is to infiltrate under the guise of reconnaissance probes. This is especially successful when the SPF are either guiding or using affiliated guerrilla forces, from a team of three to four men to a squad, or a platoon, or even a company. The guerrilla forces and/or SPF conduct small probes along the enemy defensive positions. If the enemy does not respond to these probes, the guerrilla forces and/or SPF infiltrate in small numbers and spread out. This permits larger numbers to penetrate. Once behind enemy lines, one team may cut off the escape route of the enemy, while the other units conduct a coordinated assault on both the front and flanks. The attacks will continue on all sides until the defenders are destroyed or forced to withdraw. The guerrilla forces and/or SPF will then move stealthily forward to the open flank of the next enemy position and repeat the tactics.

3-187. The OPFOR also conducts another very successful variation of this infiltration and subsequent action. Another very successful variant is that the guerrilla forces and/or SPF do not immediately attack as soon as they are successfully behind enemy lines. Once behind the enemy, they may wait a few hours or up to 3 days or more and may number as much as a full guerrilla company or even a battalion, depending on the circumstances. Once emplaced either behind the enemy or more likely behind and on both flanks of the enemy, the infiltrated force then waits for the main attack. If the main attack is successful, the enemy will either retreat or fall back. At that time, the infiltrated guerrilla forces and/or SPF will ambush and destroy the remaining enemy forces. If the main attack is faltering or appears as if it may fail, the infiltrated guerrilla forces and/or SPF simultaneously attack from both the rear and flanks, ensuring victory.

SWARMING

3-188. Swarming is a tactic that results in the convergent attack(s), from multiple directions, and possibly multiple dimensions, by numerous elements on a single target(s). Guerrillas can use this type of attack, especially when accompanied by affiliated SPF. SPF teams can plan or otherwise facilitate the attack and may or may not accompany their surrogate forces conducting the swarming attack.

3-189. There are two basic types of swarming: the massed swarm and the dispersed swarm. In the massed swarm, the elements begin as a massed (assembled) unit. On command, the elements then disassemble and conduct a convergent attack(s) to swarm the enemy from numerous directions. In the dispersed swarm, the elements are geographically dispersed from the beginning. On command, the elements infiltrate. Once prepared, they attack (from their respective directions), converging on the enemy without forming a single massed unit. Swarms are equally effective in both the offense and the defense.

3-190. Of the two types of swarming, the dispersed swarm is the most difficult to defend against because the attacking elements never present a massed target. Guerrillas prefer to use the dispersed swarm attack where the attackers are initially dispersed, then converge on the target(s). It is more appropriate to the dispersed fight that guerrillas, and affiliated SPF, prefer. Once the attack is complete, the attacking elements can either dissipate into the local population, exfiltrate back to where they came from, or move to hide positions or sanctuary (possibly cross-border).

STAY-BEHIND

3-191. Guerrilla forces, usually HK teams, may remain in areas formerly under guerrilla (or other Hybrid Threat component) control, or areas not previously occupied by the enemy. If SPF teams are also present in such areas, they can help organize the guerrilla force to conduct surveillance or direct action. Stringent precautions are taken to preserve security, particularly that of the refuge areas or other safe sites to be used during the initial period of enemy occupation. Information concerning locations and identities within the indigenous guerrilla organization is kept on a need-to-know basis. Contacts among various elements use clandestine communications. Dispersed caches, to include radio equipment, are pre-positioned when possible. SPF can provide communications, liaison, and support to guerrilla stay-behind activities.

3-192. Guerrillas (and associated SPF personnel) have a better chance of survival in small towns, villages, and rural areas. However, when stay-behind operations are attempted in heavily populated urban areas, the SPF teams may be completely dependent upon the indigenous guerrilla organization for security, the contacts required for expansion, and the buildup effort.

Chapter 4

Criminals

Criminal elements exist at every level of society and in every operational environment (OE). Their presence, whatever their level of capabilities, adds to the complexity of any OE. They may be intertwined with irregular forces and possibly with regular military and/or paramilitary forces of a nation-state. However, they may also pursue their criminal activities independent of such other actors.

CHARACTERISTICS

4-1. Some individuals, groups, and activities are criminal or illegal only because they violate laws established by a recognized governing authority. Others may violate moral or ethical standards of a given society or of the international community.

> *Note.* In some OEs, the threat is more criminal than military or paramilitary in nature. Insurgents, guerrillas, or other armed groups often use or mimic established criminal enterprises and practices to move contraband, raise funds, or otherwise further their goals and objectives.

4-2. Criminal activity is a category of violence that is enmeshed in the daily life of most people in both urban and rural areas. However, criminal activity thrives in areas where there is instability and lack of government control or law enforcement. The actions of insurgents and guerrillas further erode stability and effective governance, creating more opportunities for criminal pursuits. Sometimes, given those opportunities, insurgent or guerrillas themselves turn to crime—either to sustain themselves or for personal profit. It may be difficult to distinguish crime from ethnic feuds, ideological and theological extremism, or other elements of a culture that incite insurgency or guerrilla warfare.

4-3. Governing authorities often characterize insurgents and guerrillas as "bandits." The reason for this is that their activities in opposing the governing authority and sustaining themselves are illegal (from the government perspective). Acts of subversion may be against the law (that is, criminal) even if not violent.

SCOPE OF OPERATIONS

4-4. Criminal organizations are normally independent of nation-state control. Large-scale organizations often extend beyond national boundaries to operate regionally or worldwide. Large-scale organizations may have the capability to adversely affect legitimate political, military, and judicial organizations. However, individual criminals or small-scale criminal organizations (gangs) typically do not. Still, any criminal organization can affect such government organizations and/or military operations by becoming affiliated with the irregular OPFOR or with military forces of another nation-state.

4-5. Unless a criminal organization is in league with government officials, it must operate in ungoverned or poorly governed areas. Otherwise, the governing authority would interfere with the criminal activity. In today's world, the ungoverned area may be virtual—in the Internet and cyberspace. Criminal organizations can draw on virtual sanctuaries such as websites, chat rooms, and blogs.

4-6. Criminal organizations desire a space where they can conduct their activities unconstrained by a government. They may seek to create or maintain a region where there is no governmental control or only governmental control that they can co-opt. Such an area allows them sufficient latitude to operate and discourage rival criminal enterprises. From this base area, they can generate more and more violence and instability over wider sections of the political map.

Chapter 4

4-7. Some criminal organizations can generate instability and insecurity within a state or across borders. They can become partners with insurgents in order to further their criminal ends. A criminal organization takes on the characteristics of an insurgency when it uses subversion and violence to negate law enforcement efforts. Some criminal organizations may seek to co-opt political power through corruption and intimidation. The more they seek freedom of action, the more they inhibit state sovereignty. A criminal organization may create its own form of "government" by providing protection and enforcing its will on the populace. If it can challenge the governing authority's control beyond the local level of government, it in effect becomes an insurgency unto itself, although its ends are materially focused rather than ideological.

RELATIONS WITH OTHER ORGANIZATIONS AND ACTORS

4-8. Criminal organizations may have some type of relationship with guerrilla and/or insurgent organizations or other actors, based on similar or shared goals and/or interests. The nature of the shared goal or interest determines the tenure and type of relationship and the degree of affiliation. Any affiliation depends on the needs of the criminal organization at a particular time. To criminals, any cooperation with other actors is viewed through the lens of profitability. They may actually oppose other actors whose activities degrade their criminal enterprises.

Note. Criminals and criminal organizations, both armed and unarmed, may be considered noncombatants as long as they are neutral. However, they may be considered as combatants if they become affiliated with regular military or irregular forces. In the latter case, they can be considered part of the irregular OPFOR.

Irregular Forces

4-9. By mutual agreement, or when their interests coincide, criminal organizations may become affiliated with insurgents and/or guerrillas controlling and operating in the same area. Such allies can provide security and protection against government forces or other common enemies. They can also provide support to the criminal organization's activities. In exchange, the criminal organization may provide financial assistance, advanced technologies, or weapons.

4-10. Mutual interests of criminals, insurgents, and/or guerrillas can include preventing extraregional or local government forces from interfering in their respective spheres. The amount of mutual protection depends on the size and sophistication of each organization and its level of influence on the government or the local population.

4-11. On behalf of a criminal organization, insurgents or guerrillas can conduct—
- Diversionary actions.
- Reconnaissance and early warning.
- Money laundering.
- Smuggling.
- Transportation.
- Civic actions.

4-12. Criminal organizations may not be part of an insurgency. However, their activities—such as theft, hijackings, kidnappings, and smuggling—can further undermine the governing authority. Insurgent organizations often link themselves to criminal networks to obtain funding and logistics support. In some cases, insurgent networks and criminal networks become indistinguishable. Many insurgent organizations are more similar to organized crime in their organizational structure and relations with the populace than they are to military units.

Note. If an insurgent organization creates a shadow government, criminals may co-opt elements of that government for their own benefit or reach agreements for mutual benefit.

Nation-State Military or Paramilitary Forces

4-13. At times, criminal organizations might also be affiliated with nation-state military and/or paramilitary actors. In time of war, a state may encourage and materially support criminal actions that contribute to the breakdown of civil control in a neighboring country. Regular, reserve, and militia forces of a state can maintain various relationships with criminal organizations. Special-purpose forces (SPF) can support criminal organizations and possibly conduct operations in conjunction with them.

Local Populace

4-14. Criminal organizations may conduct civic actions to gain and maintain support of the populace. A grateful public can provide valuable security and support functions. The local citizenry may willingly provide ample intelligence collection, counterintelligence, and security support. Such support can also be the result of bribery, extortion, or coercion. For example, a criminal organization might use bribery or extortion to induce members of the local populace to act as couriers or otherwise support its activities. It might also coerce a businessperson into running a front company on its behalf.

Private Security Companies

4-15. Large criminal organizations may hire private security contractors (PSCs) to guard key facilities. PSCs can also provide bodyguards for a criminal leader or conduct surveillance or a search at a site prior to his arrival.

PERSONNEL

4-16. In most countries and cultures, criminals fall into the lower orders of social groups. However, there are some cases in which the criminals are well known and receive a degree of respect for their adherence to traditional social values and their public social works (civic actions).

4-17. Criminals may or may not be armed. For instance, individuals who perform money laundering or operate front companies might not be armed. Individual criminals or gangs might be affiliated with insurgents or guerrillas and still perform support functions that do not involve weapons. Conversely, some criminals are armed noncombatants who are not part of or affiliated with any military or paramilitary organization. Some minor criminals may use their weapons for activities such as extortion and theft. They might even steal from local government or extraregional forces, to make a profit. Opportunists may decide to hijack a vehicle or a convoy by force of arms.

4-18. Individual criminals may be completely neutral or have leanings for either side, or several sides. At least at the outset, they may not be members of or directly affiliated with the irregular OPFOR. However, since some of them are already armed, they could easily become combatants, if their situation charges. While some noncombatants are normally unarmed, there is always the potential for them to take up arms in reaction to developments in the OE and their perception of the actions of local governing authorities or an extraregional force.

> *Note.* Most criminals (especially noncombatants) are not irregular forces or part of the irregular OPFOR. However, they are included in this TC because of their relationships and interactions with the irregular OPFOR. Their criminal acts can add to the general instability caused by the irregular OPFOR.

MOTIVATIONS

4-19. Criminals, unlike insurgents or guerrillas, are seldom driven by any political or religious ideology. They exist to make money or to wield power. However, some criminal organizations may align themselves with insurgents or guerrillas if their interests are mutually supportable. Criminal organizations may even become affiliated with military or paramilitary forces of a nation-state if their interests coincide.

Chapter 4

> *Note.* Insurgents, guerrillas, or nation-states may espouse a rigid religious or political ideology. However, they may set aside those principles when using the drug trade or other criminal activities to finance their operations.

PROFIT

4-20. The primary motivation of criminals is making as much money as possible with as little risk as possible. To make a profit, they may fill a need or want with otherwise unavailable products or provide services that are illegal. To mitigate risk, they may buy influence with politicians and police officials. Failing that, they may eliminate those law enforcement and legal system personnel who interfere with their activities, as well as criminal competitors.

4-21. The enemies of criminal organizations are any political, military, legal, or judicial institutions that impede their actions and interfere with their ability to make a profit. However, there are other groups that conduct drug-trafficking or other illegal actions as a means to purchase weapons and finance other paramilitary activities. Criminals may tolerate or cooperate with such groups as long as they do not appear to be competitors or infringe on their own ability to make a profit. A criminal organization may sell or trade information to a military or paramilitary organization, for profit or mutual benefit.

POWER

4-22. Individual criminals and criminal organizations commit crimes not only for money, but also to gain power and prestige. Actions such as assassination, murder, and maiming are intended to demonstrate the criminal organization's power and control. That power may be over public officials, persons it cannot otherwise control or intimidate, criminal competitors, or people disloyal to the organization.

4-23. Larger criminal networks and transnational criminal organizations conduct criminal activities on a scale that can threaten whole societies and/or the international community. They can render governments corrupt and ineffective. They often begin to control ungoverned territory within a nation-state, acquire political power in poorly governed regions, and eventually vie with the governing authority for influence in government-controlled space. They also look for ways to exert power and influence internationally.

CRIMINAL ORGANIZATIONS

4-24. Very few crimes can be carried out by an individual criminal. Therefore, some form of organization is normally required. It may be a group of two or three individuals or a larger, more sophisticated organization. The higher the level of organization, the greater the potential for profit and power. Criminal organizations may not change their structure in wartime, unless wartime conditions favor or dictate different types of criminal action or support activities.

4-25. Weapons and equipment vary based on type and scale of criminal activity. Criminal organizations at the higher end of the scale can take on the characteristics of a paramilitary organization—either for self-protection or as a private army for hire. Criminals may have the best technology, equipment, and weapons available, simply because they have the money to buy them.

4-26. Criminal organizations generally fall into three types: gangs, large-scale criminal networks, and transnational criminal organizations. Some gangs and criminal networks develop into larger criminal networks and possibly into transnational criminal organizations. Thus, the lines of separation are not always clear-cut. However, there are some basic differences in how these three types are structured and how they operate.

GANGS

4-27. Some criminals may form loosely associated organizations with no true formal structure. These small gangs are a concern to local law enforcement, but normally are not a threat to legitimate institutions. Nevertheless, their often unpredictable acts of violence can destabilize the social and political environment. Smaller gangs tend to have loose, unsophisticated leadership focused on turf protection, gang loyalty, and local, petty crime (such as theft or small-scale drug sales). Even when many gangs with various leaders operate in the same areas, they may or may not interrelate with each other.

4-28. These smaller gangs are not irregular forces, but their impact on the local populace can be significant and may actually assist irregular forces. Even small gangs may become affiliated with insurgents or guerrillas if they stand to gain profit or power.

4-29. Some smaller gangs may develop into larger gangs or into large-scale criminal networks. This can happen in two ways. A small gang may grow into a larger organization, or several gangs may (willingly or forcibly) be joined to form a larger organization.

LARGE-SCALE CRIMINAL NETWORKS

4-30. Gangs may develop into large-scale criminal networks. At this level, they expand illicit business, increase profits, and typically focus on trafficking and market protection. These networks and the gangs that comprise them also continue with the traditional sources of illegal income. However, they are much more versatile, flexible, and able to adapt to various opportunities. The large number of criminal organizations loosely associated with one another is able to exploit more opportunities with a variety of means. Large-scale criminal networks can include any or all of the actions listed under "Criminal Activities" later in this chapter.

> *Note.* The term *criminal network* more closely captures the loose structure of many large criminal groups that do have an enduring association but are more fluid and adaptable than the more specific terms *syndicate* or *cartel* imply.

4-31. Most criminal networks are loosely structured and function primarily because each participant is pursuing his own interests. Such a network is not necessarily a formally structured, hierarchical organization with one individual controlling and running the operation. Rather, it may be a loose-knit, intricate web of individuals or groups selected for their particular skills. However, they frequently continue to operate independently until the network needs their skills. The network can quickly assemble the appropriate resources to take advantage of opportunities for specific illegal activities and then release these assets between jobs. This reduces exposure to detection and keeps individuals from having too much information about the overall operation.

4-32. Even when there are strong vertical links, there can still be a great deal of autonomy among the numerous small gangs that make up the network. The hierarchy may receive payments from those lower in the organization, but each gang is free to pursue its own criminal enterprises when the larger organization does not require its services. The small gangs may still operate relatively independently (and perhaps competitively). They do not necessarily share key information with the larger structure. Even the highest-level leadership may not be aware of all operations.

4-33. These criminal networks use violence as a means to create and protect their market as well as marginalize and control their competition. They seek to control or weaken state security institutions. They often begin to dominate community life within large areas of a nation-state. Criminal groups at this level may begin to develop overtly political agendas to improve their market share. They may overtly challenge state security and sovereignty. They may use subversion and violence as political interference to negate law enforcement efforts directed against them.

4-34. Criminal networks that control local or regional markets may have ties to and frequently do business with criminal organizations in other regions or other countries. They do so when they need wider networks of customers, fences, money-laundering expertise, access to technologies, and other essentials for an effective criminal venture.

4-35. Some criminal networks may develop into larger criminal networks or into transnational criminal organizations. In some cases, a smaller network may simply grow into a larger organization. In other cases, several networks may willingly join to form a larger organization. In still other cases, smaller organizations may be forced (by coercion or by circumstances) to become part of a larger organization.

TRANSNATIONAL CRIMINAL ORGANIZATIONS

4-36. Some criminal organizations develop into sophisticated transnational criminal organizations. These organizations may have ambitious economic and political agendas. They often begin to fill the power vacuum in ungoverned or poorly governed regions within a nation-state and to challenge government control of other regions. This provides the transnational organization with security and freedom of movement to pursue its criminal enterprises. In some cases, the organization becomes a de facto insurgency with ends focused on the material rather than ideological goals. Actions can include any or all of the items listed under "Criminal Activities" later in this chapter—such as drug and arms trafficking, money laundering, and terrorism. Transnational criminal organizations develop their own transit routes for illegal shipments and develop their own access to contraband.

4-37. Transnational criminal organizations take advantage of increased opportunities for profit and power that are found internationally. Globalization is not limited to legal trade and commerce. Criminals in various countries can cooperate in criminal ventures that take place across several countries. The increasing ease and effectiveness of global communications plays a significant role in arranging criminal ventures and in laundering the proceeds.

4-38. For example, smuggling is a big business that requires international organization. Illegal substances or legal goods less expensive elsewhere are smuggled across state boundaries. (Drugs are the most lucrative of smuggled items.) A significant part of the profits may go to suppliers and associates in other countries, and profits may be laundered using international financial systems.

4-39. Also contributing to the international nature of crime is the increased movement of people across borders. Businesses, both legitimate and illegitimate, benefit from expanding global travel. Another aspect is movement of people forced from their homes by war or political persecution. Others move in order to seek the opportunity to build a better life for themselves and their families. The vast majority of these people are not criminals. However, the size of the movement provides perfect cover for those who are connected to transnational criminal organizations. Some migrants avoid formal channels and pay smugglers to get them into another country.

COMMAND AND CONTROL

4-40. Criminal organizations (especially gangs and criminal networks) tend to have a lose structure. However, there is always some vertical link between lower-level subordinates and the organization's leadership. When smaller organizations develop into larger criminal organizations by growth or by forcible joining, the vertical links are more likely to take the form of a relatively strong hierarchical leadership structure. However, when smaller organizations willingly become part of a larger organization, there is more likely to be a rather loose leadership structure. The following paragraphs describe some of the levels that may exist even in a loose hierarchical structure.

SENIOR LEADERS

4-41. As a general rule, transnational criminal organizations tend to be highly centralized with senior leaders exerting control through a rigid chain of command. Gangs and criminal networks vary much more in their leadership structure. Some are highly organized, while others may be near anarchic with no central leadership.

4-42. Compared to gangs, the leadership of large-scale criminal networks tends to be more centralized, but with a loose hierarchical structure. However, there are exceptions. For example, some networks (such as drug cartels or crime syndicates) may be tightly organized with a rigid hierarchy. Some networks may resemble a business enterprise with specially qualified individuals on the operational side, while others supply the support services.

4-43. Different groups have varying titles for their leaders. Such titles may include boss, number one, chief, father, or simply leader (in various languages).

INTERMEDIATE LEADERS

4-44. Intermediate leaders are the "middle management" of criminal organizations. Each of these is the ranking member of his particular part of a larger organization. Again, titles vary from organization to organization. Examples are sub-boss, underboss, street boss, or franchise leader. Those intermediate leaders who, in the view of their organization, lead a crew of "soldiers," may have military-like titles (such as sergeant, lieutenant, captain, or brigadier). Regardless of the title, the mission is the same:

- Turn orders into reality and make money.
- Ensure loyalty and punish those who get out of line.
- Avoid arrest and prosecution.
- Eliminate or intimidate the competition.

SPECIALIZED AND TECHNICAL SUPPORTERS

4-45. All organizations have specialists. These are people who maintain a very high level of expertise in one or two areas necessary for the success of the organization. In criminal organizations, this traditionally includes such persons as—

- Accountants.
- Defense attorneys.
- Counselors.
- Professional assassins.
- Explosives experts.
- Counterfeiters.
- Smugglers.
- Pilots.
- Drivers.

In today's strategic and operational environments, it also includes—

- Hackers.
- Computer programmers.
- Telecommunications specialists.
- Weapons experts.
- Human traffickers.

FIELD OPERATIVES

4-46. At the lowest level of the criminal organization is the field operative. That is the man on the street, willing to commit crimes for money, power, and prestige. These people may be full members (made men), aspiring members (prospects), or those who want to be seen with the organization (hang-around, associates). The titles for field operatives may include soldier, warrior, legman, or simply member. Whatever their title, they are still criminals within a larger organization and will be the face of the organization that most of the populace see.

PASSIVE SUPPORTERS

4-47. In many cases, there are individuals or groups who provide passive support to criminal organizations for any number of reasons. Some see the organization as a defense against another group or corrupt government officials. Some have familial connections. Some see neighborhood, community, or country loyalty embodied in the criminal organization. Some are merely concerned about their own financial security or material gain.

CRIMINAL ACTIVITIES

4-48. Criminals use many and varied tactics and techniques. Some of these methods overlap with one another. The activities typically include an objective to make fiscal profit and/or achieve influence.

SECURITY

4-49. Security is crucial for criminal organizations. They may use the highest degree of sophistication available to conduct intelligence collection and counterintelligence activities. These activities are a priority and can be well funded. Intelligence sources may extend to high levels within government and law-enforcement agencies. The local populace may willingly provide ample intelligence collection, counterintelligence, and security support. Intelligence and security can also be the result of bribery, extortion, or coercion.

4-50. Most members of criminal organizations are capable of protecting themselves and their assets. Typically, they carry small-caliber weapons, such as handguns, pistols, rifles, and shotguns. They are lightly armed out of necessity or convenience, not for lack of resources. When greater force of arms is necessary to control people, protect vital resources, or obtain information, these organizations typically have members who can use heavier arms, such as machineguns and assault weapons. Large criminal organizations may hire PSCs to conduct surveillance, provide personal security for leaders, or guard key facilities.

THEFT

4-51. Theft is the taking of another person's property without that person's permission or consent with the intent to deprive the rightful owner of it. Thus, *theft* is an overarching term that covers various crimes against property such as burglary, embezzlement, larceny, looting, robbery, and fraud. (See also identity theft and intellectual property theft, both under Cyber Crime.)

FRAUD

4-52. A fraud is an intentional deception made for personal gain and/or to damage another individual or entity. Defrauding people or entities of money or valuables is a common purpose of fraud. Fraud can include—

- Ponzi schemes.
- Insider trading.
- Embezzlement.
- Insurance scams.
- Money laundering.
- Forgery.

Fraud can also include cyber crimes such as—

- Identity theft.
- Copyright infringement.

RACKETEERING

4-53. A racket is a fraudulent scheme, enterprise, or activity. It is usually an illegitimate business made workable by bribery or intimidation. A racketeer is one who extorts money or advantages by threats of violence, by blackmail, or by unlawful interference with business or employment. Therefore, racketeering overlaps with bribery, intimidation, and extortion.

4-54. In racketeering, a criminal organization typically creates or perpetuates a problem or the perception of a problem for which it then offers a solution, for a fee. The intent is to engender continual patronage. In the traditional example of a protection racket, the racketeer informs a store owner that a substantial monthly fee will be required in exchange for protection. The "protection" provided takes the form of the absence of damage inflicted upon the store or its employees by the racket itself. Another example is malicious software (malware) that pretends to be detecting spyware or other infections on a computer and offers to download a cleaning utility for a fee. In actuality, the distributor of the malware is also the maker of the cleaning utility. (This aspect of racketeering overlaps with the category of cyber crime, discussed below.) In addition to protection rackets, racketeering can also involve numbers rackets or illegal lottery.

4-55. Racketeering can also include seemingly legitimate businesses that provide a front for illegal activity such as buying and selling illegal merchandise. Loan sharking is another form of racketeering in which a criminal offers loans at extremely high interest rates to borrowers who cannot qualify for loans from legitimate sources. The debtor faces violence or other criminal means to cause harm if the debt is not paid.

4-56. Black markets fall under racketeering as illegal businesses. These can include traded goods and services such as biological organs, transportation providers, illegal drugs, prostitution, weaponry, alcohol, tobacco, currency, and fuel. Black markets tend to flourish in most countries during wartime, due to rationing, and impact commodities such as food, fuel, rubber, and metal.

GAMBLING

4-57. In most societies, the act of gambling is in itself not illegal. What becomes illegal is an illicit business based on gambling (also called gaming). It is illegal because either—
- It is conducted in a fraudulent manner.
- The governing authority has outlawed it.
- The gambling enterprise fails to share all its profits with the governing authority.

Gambling in the form of a numbers racket or illegal lottery is a form of racketeering. Internet gambling is a form of cyber crime.

4-58. Many types of gambling have been, or still are, illegal in some places. Hence, criminals may be the only operators of some games. Even when a governing authority legalizes some gambling ventures, it may be difficult to keep criminals from becoming involved, due to the huge potential for profits. Besides gaming operations recognized and (sometimes poorly) regulated by the state, there are many other gambling opportunities in which criminals may be involved.

4-59. Gambling is often associated with other types of crime. Operators of legal gambling establishments may receive kickbacks for allowing money laundering. Gambling operations earn large amounts of cash and present particular opportunities for skimming and money laundering. Aside from skimming profits from a legalized gambling enterprise, skimming can also occur with the granting of credit, which can lead into loan sharking. In particular, problem gamblers who need money for gambling may easily fall prey to loan sharks. Compulsive gambling can also lead to crimes such as embezzlement, robbery, check forgery, and fraud.

PROSTITUTION

4-60. Prostitution is the practice of indulging in sexual relations for money. As a criminal enterprise, it also involves a criminal or criminal organization that profits from the commercial sex act. Prostitutes may also profit from the venture, although the criminal organization gets its cut. However, prostitution may be linked to human trafficking, in which case the criminals get all the profits.

EXTORTION

4-61. Extortion is the act of obtaining money, materiel, information, or support by force or intimidation. Criminal organizations use extortion to obtain information or cooperation or to protect members. Examples of extortion include—
- Intimidating politicians to vote in a manner favorable to the criminal organization.
- Intimidating judges to free an organization member.
- Forcing a farmer to grow drug-producing crops.
- Extorting money from local businesses in exchange for protection, which means not harming the business or its members.

- Using death threats against an individual or his family to cause him to provide information or resources.
- Intimidating other people not to take action against the criminal organization.
- Using information warfare (INFOWAR) methods to create and maintain fear caused by extortion.

BRIBERY

4-62. Bribery is giving money or other favors to influence someone. Criminals give money to people in power who make or influence decisions. For example, bribes to law-enforcement officials can cause them to have their patrols avoid a criminal organization's transit routes. If the organization is unable to bribe someone, it employs harsher methods, such as extortion, assassination, or murder, to gain cooperation.

ARSON

4-63. Arson is maliciously burning another person's dwelling, structures, or property. This may be done to as punishment for noncompliance with internal rules of criminal organizations. It may be the result of not paying the criminal organization for "protection." It may also be done on a for-hire basis or as a means to inflict terror on a targeted person or group.

HIJACKING

4-64. Hijacking is stealing or commandeering a conveyance. Criminals may conduct a hijacking to produce a spectacular hostage situation. Sometimes criminals may hijack a conveyance as a means of escape; in that case, the criminals may eliminate any unneeded people and materiel, such as hostages or baggage.

KIDNAPPING

4-65. Kidnapping is an abduction or transportation of a person or group by force. The person is kept in false imprisonment (confinement without legal authority). This may be done for ransom or in furtherance of another crime (such as human trafficking or hostage taking). This type of crime has become very popular with criminal organizations, and the methods vary by region. Kidnapping flourishes particularly in fragile or failed states and regions in conflict, as drug traffickers and other criminal organizations fill the vacuum left by governing authority.

4-66. The risk in kidnapping is relatively lower than in hostage taking. This is primarily because the criminals take the kidnapped victim to a location controlled by the criminal organization. The criminals then make demands and are willing to hold a victim for a long time, if necessary.

HOSTAGE TAKING

4-67. Hostage taking is typically an overt seizure of a person or persons to gain publicity, concessions, or ransom. Unlike kidnapping, where usually a prominent individual is taken, the hostages are usually not well known figures. Criminals attempt to hold hostages in a neutral or friendly area. The planning and execution of a hostage taking are similar to those of a kidnapping or hijacking. However, criminals may also take hostages as an expedient measure when they have difficulty exiting a crime scene.

MURDER

4-68. Murder is the unlawful killing of another human being without justification or excuse. Murder is perhaps the single most serious criminal offense. Criminal organizations use murder as an enforcement tool and as a method of generating revenue in murder-for-hire schemes.

ASSASSINATION

4-69. An assassination is the murder (usually of a prominent person) by a sudden and/or secret attack. It is usually prompted by religious, ideological, political, or military motives, but may be done for payment. Criminals and criminal organizations may use an assassination—

- For monetary gain.
- To exert terror.
- To display power.
- To exact revenge on a public official.
- To eliminate people they cannot intimidate.
- To punish people who have left the criminal organization.

MAIMING

4-70. Maiming is a deliberate act to mutilate, disfigure, or severely wound a person so as to cause lasting damage. Maiming can involve assault and battery on a person with intent to inflict serious injury. Other methods include—
- Male castration.
- Female genital mutilation.
- Burning or branding.
- Forcible tattooing.
- Cutting off limbs.
- Removal of tongue, eyes, or ears.
- Throwing a corrosive acid or alkali to cause blindness or scarring.

4-71. The person maimed is an outward sign of the criminal organization's power and control. A criminal organization often uses or threatens maiming—
- To enforce order within the organization.
- To collect debts.
- For extortion.
- As a for-hire operation to generate revenue.

SMUGGLING

4-72. Smuggling is the clandestine transportation of illegal goods or persons. It usually involves illegal movement across an international border. There are various motivations to smuggle. These include participation in illegal trade, illegal immigration or emigration, and tax evasion. Smuggling is often related to trafficking in persons, drugs, or arms.

TRAFFICKING

4-73. Trafficking is the transportation of goods or persons for the purpose of making a profit. Criminals conduct illegal trafficking. Human trafficking (trafficking in persons) is the second largest criminal activity in the world—followed by drug trafficking and arms trafficking.

Humans

4-74. Human trafficking is the recruitment, transportation, transfer, harboring, or receipt of persons (including children) for the purpose of exploitation. It involves the threat or use of force, coercion, abduction, fraud, deception, and/or abuse. Human trafficking may take two forms:
- Sex trafficking in which a person is induced to perform a commercial sex act.
- The recruitment, harboring, transportation, provision, or obtaining of a person for labor or services, for the purpose of subjection to involuntary servitude, debt bondage, or slavery.

Criminals choose to traffic human beings because, unlike other commodities, people can be used repeatedly and because human trafficking requires little in terms of capital investment.

4-75. Human trafficking is not the same as people smuggling. A smuggler may facilitate illegal entry into a country for a fee, but on arrival at their destination, the smuggled person is free. The trafficking victim is coerced in some way and is further exploited after arrival, in order to derive profits. Victims do not agree to be trafficked; they are tricked, lured by false promises, or forced into it. Traffickers control their victims by coercive tactics including deception, fraud, intimidation, isolation, physical threats and use of force, debt bondage, or even force-feeding drugs.

Drugs

4-76. The illegal drug trade is a black market consisting of production, distribution, packaging, and sale of illegal psychoactive substances. The legality or illegality of the black markets purveying the drug trade is relative to geographic location. The drug-producing countries may be inclined to tolerate the drug traffickers because of bribery or the effect on the country's economy. Drugs often cross international borders in order to reach the best paying customers. The massive profits inherent to the drug trade serve to extend its reach. The social consequences of drug trade include crime, violence, and social unrest.

Arms

4-77. Arms trafficking involves illicit transfers of arms, ammunition, and associated materials. Criminal organizations may be involved in two types of arms trafficking:
- Small-scale transactions by individuals or small firms that deliberately transfer arms to illicit recipients.
- Higher-value or more difficult illicit shipments of arms involving corrupt officials, brokers, or middle men motivated mainly by profit.

4-78. Arms trafficking is driven by a variety of clients, which include—
- Embargoed governments.
- Armed groups involved in war, banditry, terrorism, or insurgency.
- Criminals and criminal organizations.
- Citizens who cannot obtain guns legally.

4-79. Some arms and ammunition may come from illicit arms manufacturers. However, the source of a large proportion of illicit conventional arms is government disposals of surplus arms or thefts from insecure government stockpiles. Governments themselves may deliberately facilitate covert flows of arms to their proxies or allies, or to embargoed or suspect destinations for profit.

4-80. Arms trafficking is widespread in regions of political turmoil. However, it is not limited to such areas. Most arms trafficking occurs at the regional or local level. Among the most common forms are numerous shipments of small numbers of weapons that, over time, result in the accumulation of large numbers of illicit weapons. While individual transactions occur on a small scale and do not draw attention, the sum total of weapons trafficked is large.

CYBER CRIME

4-81. Cyber crimes are offenses targeting or using information technology. This includes computers, computer networks, and other telecommunication networks such as the Internet (chat rooms, emails, notice boards, and groups) and mobile phones. Such crimes may threaten not only individuals and groups but also a nation's security and financial health. Cyber crimes can facilitate a variety of other criminal activities, including money laundering, extortion, fraud, racketeering, gambling, smuggling, and trafficking.

4-82. Criminals exploit the speed, convenience, and anonymity that modern technologies offer in order to commit a diverse range of criminal activities. The global nature of the Internet allows criminals to commit almost any online illegal activity anywhere in the world. In the past, cyber crime has been committed by individuals or small groups of individuals. However, an emerging trend is for criminal networks and criminally minded technology professionals to work together and pool their resources and expertise.

4-83. Cyber crimes include network intrusions, hacking attacks, malicious software, and account takeovers leading to significant data breaches affecting every sector of the world economy. Advances in computer and telecommunications technology and greater access to personal information via the Internet have created a virtual marketplace for transnational cyber criminals to share stolen information and criminal methodologies. The increasing level of collaboration among cyber criminals raises the level of potential

harm to individuals, companies, and governments. Members of online forums discuss cyber crime topics of interest. Criminal purveyors buy, sell, and trade—
- Malicious software.
- Hacking services.
- Spamming devices.
- Personal identification information.
- Credit and debit card data.
- Bank account information.
- Brokerage account information.
- Counterfeit identity documents.
- Other forms of contraband.

4-84. Cyber crime falls into two broad categories:
- Computer crimes.
- Intellectual property crimes.

Computer Crimes

4-85. Computer crime refers to any crime that involves a computer or computer network. The computer or network may be the target, or it may be used in the commission of a crime, the primary target of which is independent of the computer network or device.

Crimes Targeting Computers

4-86. Criminals can cause damage to computers in many ways. For example, an unauthorized intruder can send commands that delete files or shut the computer down. Intruders can initiate a denial-of-service attack that floods the victim computer with useless information and prevents legitimate users from accessing it. A virus or worm can use up all of the available communications bandwidth on an agency or corporate network, making it unavailable to employees. When a virus or worm penetrates a computer's security, it can delete files, crash the computer, install malicious software, or do other things that impair the computer's integrity.

4-87. Extortion threats involving damage to a computer are a high-technology variation of old-fashioned extortion. Intruders may threaten, unless their demands are met, to—
- Penetrate a system and encrypt or delete a database (erasing or corrupting data or programs).
- Distribute denial-of-service attacks that would shut down (or slow down) the victim's computers.
- Steal confidential data.

The threat need not be sent electronically.

4-88. Crimes may involve intercepting or interfering with communications through the use of electronic, mechanical, or other devices. This applies to electronic communications as well as oral and wire communications via common carrier transmissions. This may involve—
- Spyware or intruders using packet sniffers.
- Persons improperly cloning email accounts.
- Other surreptitious collection of communications from a victim's computer.

The crime may extend to the unauthorized disclosure of the contents of an illegally intercepted communication or using such information for other criminal purposes. Such crimes may also injure or destroy various types of communication operated or controlled by a government or used for a state's military or paramilitary functions.

Crimes Using Computers

4-89. Crimes using computers or facilitated by computers include—
- Unauthorized access to computers (even government computers and perhaps law enforcement).

Chapter 4

- Unauthorized access to stored communications (including email, social networking data, and voicemail).

Two of the most common uses are for—
- Fraud (computer fraud and wire fraud).
- Identity theft.

4-90. **Fraud.** Computer fraud is the use of information technology to commit fraud. Criminals may devise various schemes or artifices to obtain money, property, goods, or services of measurable value. These include—
- Accessing a computer without authorization to obtain—
    - Commercial data.
    - Information contained in a financial record of a financial institution.
    - Information contained in a file of a consumer reporting agency on a customer.
    - Information from any government department or agency.
    - Information from a protected computer.
- Accessing without authorization a government, commercial, or personal computer and affecting the use of the owner's operation of the computer.
- Accessing a protected computer with the intent to defraud and thereby obtain anything of value.
- Causing the transmission of a program, information, code, or command that causes damage to a computer system, personal injury, or a threat to public health or safety.
- Trafficking in passwords or similar information through which a computer may be accessed without authorization.
- Manipulating market data for criminal purposes.
- Data or information or even the use of a computer can be regarded as a thing of value.

4-91. Computer fraud can also fall under the more general category of wire fraud. In what is considered wire fraud, criminals may use fraudulent pretenses, representations, or promises transmitted by various means of telecommunication, including—
- Wire.
- Radio.
- Television.
- Mobile phone.
- Facsimile.
- Telex.
- Modem.
- Internet.

4-92. Money mule schemes could be categorized as computer or wire fraud. In such schemes, criminals trick people into moving money for them. Through Internet phishing, criminals can steal money from unsuspecting people by accessing their accounts. The phishers then face the problem of moving the large sums of money acquired from the victims to their own accounts, often in other countries, without attracting suspicion. Traditionally, drug smugglers have used people (referred to as mules) willing to carry small amounts of drugs across borders for them for a price. In a variant of this scheme, phishers recruit innocent people in the countries where their victims reside by offering them a lucrative job using their home computers. Those recruited are usually unaware that these are not legitimate business opportunities. Thus, phishers can dupe a large number of individuals (called money mules) into accepting relatively small amounts of money stolen from people's accounts and transferring the funds to the phishers in return for a commission.

4-93. **Identity Theft.** Identity theft is almost always committed to facilitate other crimes. It involves unauthorized transfer, possession, or use of a means of identification of another person with the intent to commit, or to aid or abet, or in connection with, any unlawful activity. Most commonly, it involves the misuse of another individual's personal identifying information for fraudulent purposes. With relatively little effort, an identity thief can use this information to take over existing credit accounts, create new accounts in the victim's name, or even evade law enforcement after the commission of a violent crime. Identity thieves also sell personal information online to the highest bidder, often resulting in the stolen information being used by a number of different perpetrators. Personal information can also be obtained for the purpose of blackmail or other forms of extortion.

4-94. Criminals can obtain credit and debit card numbers by hacking into the wireless computer networks of major retailers. Network intrusions can compromise the privacy of individuals if data about them or their transactions resides on the victim network. Once inside the networks, criminals install "sniffer" programs that capture card numbers, as well as passwords and account information. After collecting the data, they conceal it in encrypted computer servers that they control. Then they sell the credit and debit card numbers through online transactions to other criminals. In addition to these "carders," there are "phishers," who obtain the same type of information via fraudulent emails.

4-95. The unlawful use of identification information can involve access device fraud. Such fraud can include any device, number, or other means of account access that can be used to obtain money, goods, services, or other things of value. It can also involve unlawful access to buildings or facilities.

Intellectual Property Crimes

4-96. Intellectual property crimes involve theft of material protected by copyright, trademark, patent, or trade-secret designation. Such intellectual property is vital to local, national, and international economies.

4-97. The interconnected global economy creates unprecedented business opportunities to market and sell intellectual property worldwide. Geographic borders present no impediment to international distribution channels. If the product cannot be immediately downloaded to a computer, it can be shipped and arrive by next day air. However, the same technology that benefits rights-holders and consumers also benefits intellectual property thieves seeking to make a fast, low-risk profit. In addition, trafficking in counterfeit merchandise also generates large profits.

4-98. The most egregious violators are large-scale criminal networks and transnational criminal organizations whose conduct threatens not only intellectual property owners but also the economy of nation-states. Because many violations of intellectual property rights involve no loss of tangible property and do not require direct contact with the rights-holder, the owner often does not even know that it is a victim for some time.

4-99. Intellectual property crimes can overlap with computer crimes, especially in the following areas:
- Unauthorized access of a computer to obtain information.
- Mail or wire fraud (can include the Internet).
- Devices to intercept communications.

4-100. Unauthorized obtainment of information or electronic media covered by copyright, trademark, patent, or trade-secret designation robs the rights-holders of their ideas, inventions, and creative expressions. Such theft is facilitated by digital technologies and Internet file sharing networks.

4-101. Although fraud schemes can involve copyrighted works, it is not mail or wire fraud unless there is evidence of any misrepresentation or scheme to defraud. Mail and wire fraud may exist, even if the perpetrator tells his direct purchasers that his goods were counterfeit, as long as he and his direct purchasers intended to defraud the direct purchaser's customers. Wire fraud can include the Internet.

4-102. Intellectual property crimes may involve intercepting and acquiring the contents of communications through the use of electronic, mechanical, or other devices. Criminals may then market the contents of an illegally intercepted communication, including intellectual property.

MONEY LAUNDERING

4-103. A criminal organization conducts money-laundering activities to transfer funds into the legitimate international financial system. Because of the legal restrictions levied by a governing authority, the organization must have a way of transferring "dirty" (illegally earned) money into "clean" money. The organization smuggles some currency back to its country of origin. However, large sums of foreign currency are not feasible for the organization because it must use the legal currency of its country to make transactions. For example, a farmer cannot receive payment for his drug crops in the foreign currency because he would be unable to use it.

4-104. A more productive way to transfer funds into the legitimate financial system is to operate through front companies. Front companies are legitimate businesses that provide a means to launder money. The criminal organization establishes its own front companies or approaches legal companies to act as intermediaries. Some front companies, such as an import or export business, may operate for the sole purpose of laundering money. Other companies (possibly targets of extortion) operate as profit-making activities and launder money as a service to the organization. Criminal organizations operate front companies in their own country, as well as in other countries.

4-105. Personnel involved in money laundering can include accountants, bankers, tellers, and couriers. Some are willing participants in the money-laundering process and accept bribes for their services. Extortion and intimidation keep unwilling members active in the process. Members of money-laundering organizations also establish and operate front companies.

4-106. Some members, such as accountants and bankers, perform their normal functions as in a legal business. However, they may conduct illegal acts on behalf of the organization. Because of banking regulations, members must conduct activities that do not draw attention to themselves. As an example, couriers deposit in bank accounts a level of funds small enough to avoid banking regulations. However, not all countries have such regulations, making it easier to launder large sums of money.

4-107. Couriers conduct many transactions with financial institutions. They travel from bank to bank making deposits or converting money into checks and money orders to prepare for smuggling activities. Money-laundering personnel may also use electronic fund transfers, which make tracking of illegal transactions much more difficult, especially when the country has lax banking laws.

CIVIC ACTIONS

4-108. Criminal organizations conduct programs of patronage under the guise of civic actions. These programs are only indirectly intended to benefit the general populace. Rather, the main intent of the criminal organizations is to gain and maintain support, reward their supporters, and facilitate their continued activities. They may build a school, improve a road, or supply food and medicine. These projects benefit the local population because they improve the people's quality of life and may improve their standard of living. For example, road building makes it easier to transport goods to market and creates jobs. Some of the jobs may be temporary, such as those in construction, while others may be permanent, such as those in education and clinics. Through its propaganda efforts, the criminal organization ensures that the population knows who is making the improvements.

4-109. Sometimes the criminal organization cooperates with an insurgent force to conduct civic actions. They do not cooperate for ideological reasons. Their sole purpose is to inhibit the governing authority's ability to affect their activities by gaining the inherent security a grateful public can provide.

INFORMATION WARFARE

4-110. Criminal organizations may have the resources to conduct a variety of information warfare (INFOWAR) activities (see appendix A). However, their focus is often on a well-orchestrated perception management (propaganda) effort. This can be a powerful tool for intimidating enemies (governing authority and criminal competitors) and encouraging support of the organization's money-making efforts. The use of perception management techniques can ensure that the population knows who is making improvements in the local environment and allow the organization to take credit for other benefits it provides to the local population.

4-111. The criminal organization can also use counterpropaganda to spin events against the governing authority. For example, the government may burn a farmer's drug-producing crop and then broadcast announcements that the action is a direct result of the farmer's involvement with an illegal activity. However, the criminal organization can counter this by instilling the idea that the governing authority does not have programs to help the farmer make profits legally and reminding the populace of the new medical clinic it built.

4-112. Various elements of INFOWAR can also be instrumental in other criminal activities. For instance, perception management and deception are related to fraud, including computer fraud. Information attack and computer warfare are parts of cyber crime.

TERRORISM

4-113. Terrorism is always illegal, whether conducted by insurgents, guerrillas, or criminals. When conducted by regular military forces or other agencies of a nation-state (perhaps in the form of ethnic cleansing or confiscation of private property), other states can consider it illegal by international standards. For criminal organizations, terrorism becomes an action against an adverse governing authority, a criminal competitor, or an innocent populace. (See chapter 6 for more information on terrorism.)

> *Note.* Many of the tactics and techniques used by criminals and criminal organizations are the same as addressed for terrorism in chapter 6. The only difference is that criminals may use these as means to obtain profits rather than just to instill fear or coerce governments and societies.

This page intentionally left blank.

Chapter 5

Noncombatants

A host of noncombatants add complexity to any operational environment (OE). The irregular OPFOR attempts to manipulate these noncombatants in ways that support its goals and objectives. Many noncombatants are completely innocent of any involvement with the irregular OPFOR. However, the irregular OPFOR will seek the advantage of operating within a relevant population of noncombatants whose allegiance and/or support it can sway in its favor. This can include clandestine yet willing active support (as combatants), as well as coerced support, support through passive or sympathetic measures, and/or unknowing or unwitting support by noncombatants.

GENERAL CHARACTERISTICS

5-1. Noncombatants are persons not actively participating in combat or actively supporting of any of the forces involved in combat. They can be either armed or unarmed. Figure 5-1 shows examples of these two basic types of noncombatants that can be manipulated by the irregular OPFOR. These examples are not all-inclusive, and some of the example entities can be either armed or unarmed. (See Armed Combatants and Unarmed Combatants later in this chapter for more detail on various subcategories.)

> *Note.* From a U.S. viewpoint, the status of noncombatants is typically friendly, neutral, or unknown. Conversely, the noncombatants would view U.S. and/or local governing authority forces as friendly or neutral in regard to themselves. For the sake of consistency throughout the chapters of this TC, however, this chapter occasionally refers to the governing authority and associated U.S. or coalition forces as "enemy," referring to the enemy of the irregular OPFOR.

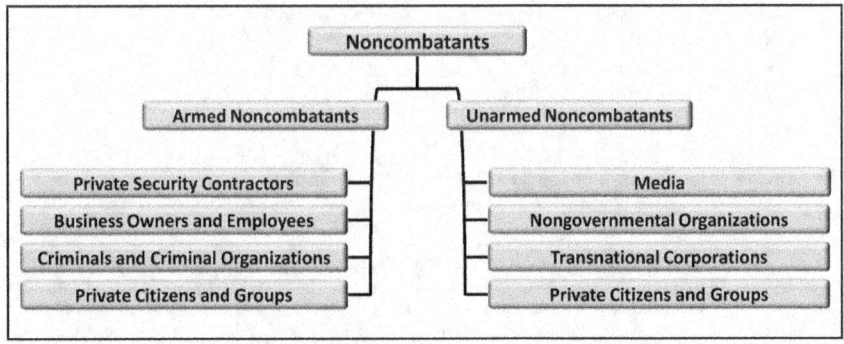

Figure 5-1. Armed and unarmed noncombatants (examples)

5-2. Aside from military and paramilitary forces, the civilian population of a nation or region is often the single most important aspect of an OE. This situation can be further complicated by the presence of other noncombatants who are not indigenous to the country or region.

RELATION TO THE IRREGULAR OPFOR

5-3. The irregular OPFOR recognizes that noncombatants living and/or working in an area of conflict can be a significant source of—
- Intelligence collection.
- Reconnaissance and surveillance.
- Technical skills.
- General logistics support.

Therefore, the irregular OPFOR actively uses noncombatants within a relevant population to support its goals and objectives. It sees them as a potential multiplier of irregular OPFOR effectiveness. It will also attempt to use the presence of noncombatants to limit the effectiveness of its enemies. The irregular OPFOR can marshal and conceal its combatant capabilities while hiding among armed and unarmed noncombatants in a geographic or cyber location.

5-4. Changes in an OE and persuasion by the irregular OPFOR's information warfare (INFOWAR) activities can cause rapid or gradual shifts in allegiance by noncombatants. Any noncombatant is a potential recruit to become a combatant in support of irregular OPFOR goals and objectives. However, the irregular OPFOR will also attempt to use those who remain noncombatants to its advantage. It purposely exploits noncombatants to cause doubt or hesitation in an enemy's decision to act. It takes advantage of the difficulties its enemies may have in distinguishing combatants from noncombatants. This uncertainty exists because both may or may not be armed and may appear to not be participating in or supporting irregular OPFOR actions. The irregular OPFOR may use noncombatants for shielding its operations close to and/or within a relevant population, where enemy reaction may cause harm to noncombatants. Incidents such as noncombatant detention and/or casualties can alienate a relevant population from enemy forces.

> *Note.* The Geneva Conventions state criteria for combatants as individuals who operate under the recognized command and control of an organization, are clearly armed, and do not attempt to disguise their intended actions when deploying for and conducting military-like actions. The irregular OPFOR is not bound by such conventions. The U.S. Army and regulated military forces of other nations consider chaplains, medical doctors, and medics as noncombatants. However, the irregular OPFOR does not necessarily recognize such functional distinctions. In guerrilla units, for example, medics are combatants, trained to fight alongside other guerrillas.

MOTIVATIONS

5-5. The irregular OPFOR uses several forms of persuasion to motivate and obtain support from noncombatants in a relevant population. Means of persuasion can include—
- INFOWAR activities.
- Social welfare programs.
- Political activism and mobilization.
- Coercion.

5-6. When co-opted directly or indirectly by the irregular OPFOR, noncombatants are a means to weaken control and legitimacy of a governing authority over its population. The irregular OPFOR attempts to communicate a compelling narrative of its own legitimacy that is accepted by the relevant population. Its visible actions, often localized in perspective, focus on demonstrating its power and authority.

5-7. The irregular OPFOR may appeal to noncombatants based on unresolved grievances of a relevant population. These unresolved grievances, perceived or factual, create conditions where individuals believe they must act to obtain a just solution. The irregular OPFOR can also appeal to aspects of ethnicity, geography, and regional history that affect personal and group relationships. These are reflected in social status and networks, lifestyle, employment, religion, and politics. The irregular OPFOR can manipulate at least three areas of grievance by to obtain noncombatant support:
- Personal or social identity and/or social mobility or advancement.
- Religious beliefs and/or persecution.

- Unjust political representation and/or governance.

5-8. Motivations to support the irregular OPFOR may vary from religious extremism to pure criminality for personal or organizational profit. Often, the deciding factor may be the desire for local freedom from control by the governing authority and its international supporters.

TYPES OF SUPPORT

5-9. The relevant population can willingly provide active or passive support to the irregular OPFOR. If necessary, however, noncombatants can be coerced by the irregular OPFOR. Some noncombatants may be aware of irregular OPFOR activities and choose to remain passive and not report information to the governing authority. Noncombatants may be sympathetic to the irregular OPFOR but remain uninvolved in any activity. Other noncombatants may unknowingly support irregular OPFOR initiatives such as money donations to charities or apparent humanitarian relief organizations that are actually front organizations for irregular OPFOR financing and/or materiel support. Those members of the local populace who elect to participate in or actively support the irregular OPFOR become combatants, even if they do not bear arms. (See Unarmed Combatants at the end of this chapter.) Figure 5-2 shows various types of support to the irregular OPFOR.

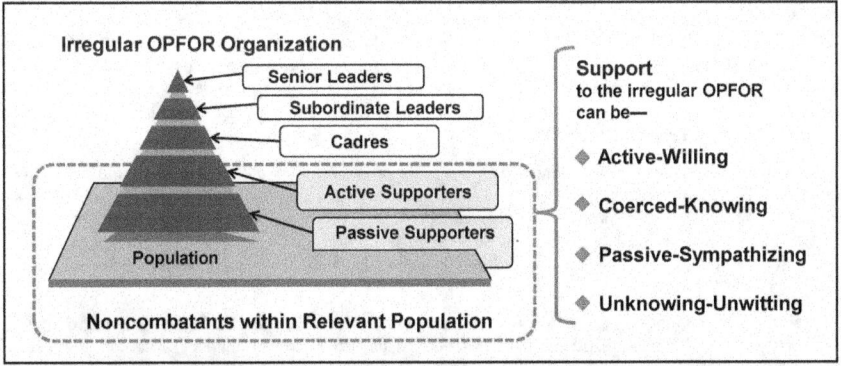

Figure 5-2. Noncombatant support to the irregular OPFOR (examples)

ARMED NONCOMBATANTS

5-10. In any OE, there are likely to be nonmilitary personnel who are armed but are not part of an organized paramilitary or military structure. Nonetheless, such people may be disgruntled and hostile to the governing authority or forces that support it. Armed noncombatants may represent a large portion of the undecided citizens in a population. Some of these nonaffiliated people may possess small arms legally to protect their families, homes, and/or businesses. Some may use weapons as part of their occupation (such as hunters, security guards, or local police). Some may be minor criminals who use their weapons for activities such as theft or extortion. Given the fact that they are already armed, it would be easy for such noncombatants to become combatants. Any number of reasons, including prejudices and grievances, can cause them to choose sides or change sides. They may switch allegiances repeatedly as circumstances evolve.

5-11. Some armed noncombatant entities can be completely legitimate enterprises. However, some activities can be criminal under the guise of legitimate business. The irregular OPFOR can embed operatives in legitimate commercial enterprises or criminal activities to obtain information and/or capabilities not otherwise available to it. Actions of such operatives can include sabotage of selected commodities and/or services. They may also co-opt capabilities of a governing authority infrastructure and civil enterprises to support irregular OPFOR operations.

5-12. Examples of armed noncombatants commonly operating in an OE are—
- Private security contractor (PSC) organizations.
- Local business owners and employees.
- Private citizens and private groups authorized to carry and use weapons.
- Criminals and/or organizations with labels such as cartels, gangs.
- Ad hoc local "militia" or neighborhood watch programs.

PRIVATE SECURITY CONTRACTORS

5-13. Private security contractors (PSCs) are commercial business enterprises that provide security and related services on a contractual basis. PSCs are employed to prevent, detect, and counter intrusions or theft; protect property and people; enforce rules and regulations; and conduct investigations. They may also be used to neutralize any real or perceived threat. PSCs can act as an adjunct to other security measures and provide advisors, instructors, and support and services personnel for a state's military, paramilitary, and police forces. They may also be employed by private individuals and businesses (including transnational corporations).

5-14. PSCs may be legitimate, well-respected corporations providing contract advisors and employees as part of a military nation-building program funded by a foreign government. A PSC that provides services on a contract basis outside its country of origin also falls into the category of a transnational corporation. Other PSCs may be domestic firms that supply contract guard forces. In its simplest form, a PSC might be a local citizen organization that performs actions on a short-term contractual basis.

Relation to the Irregular OPFOR

5-15. The irregular OPFOR can identify required or desired capabilities in a PSC and either overtly or covertly obtain the capabilities. It can also infiltrate a PSC with its own operatives, to covertly pursue its own objectives. A PSC's use of local hires can make it relatively easy to infiltrate with actual or forged documentation. Infiltration can allow the irregular OPFOR to observe actions and procedures, collect information and intelligence for future sabotage, and/or incite dissatisfaction within and among foreign and local contractors.

5-16. The irregular OPFOR might actually be a customer of a PSC. For example, the leader of an insurgent or criminal organization may employ a PSC to provide bodyguards or conduct surveillance or a search and assessment prior to his arrival. Another group, such as a drug organization, may contract a PSC to guard its facilities. (A drug organization can afford to pay more than many small countries.) The PSC may erect a fence or employ roving, armed guards to protect facilities.

5-17. PSCs use both active and passive measures that may be rudimentary or employ advanced technology such as sophisticated surveillance, identification devices, and alarms. During the conduct of their duties, members of a PSC can take offensive actions. For example, a security, guard, or patrol team might conduct an ambush to counter an intrusion by a governing authority on an irregular OPFOR operation.

5-18. The irregular OPFOR's INFOWAR activities can spotlight known or alleged abuses by PSCs in order to build negative sentiments against governing authority and/or coalition actions. Incidents can include PSCs—
- Shooting civilians misidentified as combatants.
- Using excessive force.
- Being insensitive to local customs and beliefs.
- Other inappropriate actions.

The irregular OPFOR can exploit such incidents to damage international and coalition relationships and trust. Additional public dissent might limit use of PSCs or restrict where and how they can be used. Prohibition on use of PSCs in certain circumstances could further constrain governing authority and coalition programs. It could also place additional operational requirements on enemy military forces.

Functions

5-19. Most functions of a PSC involve protecting personnel, facilities, or activities. Such security functions normally require armed contractors. However, PSCs can also perform other, security-related functions that do not require armed personnel.

Armed Functions

5-20. Functions typically requiring armed personnel can include—
- Personal security details to protect a person or group of people.
- Guard protection of static sites (such as housing areas, building sites, government complexes, and businesses—both legal and illegal).
- Transport security support to convoys and special materiel shipments.
- Security escorts.
- Cash transport.
- Covert operations.
- Surveillance.
- Intelligence services.
- Advising and/or training of indigenous or extraregional security forces.
- Operations and administration within governing authority prisons and/or detention facilities.

Unarmed Functions

5-21. Functions typically not requiring armed personnel include—
- Unarmed security functions when presence is deemed an appropriate deterrent.
- Air surveillance.
- Psychological warfare.
- Intelligence support (including information collection and threat analysis).
- Operational coordination (such as command and control, management, and communications).
- Personnel and budget vetting.
- Hostage negotiation services.
- Risk advisory services.
- Weapons procurement.
- Weapons destruction.
- Transportation support.

Advantages and Disadvantages

5-22. PSCs provide key capabilities and can often be hired quickly and deployed faster than a military force with similar skill sets. This flexibility allows governmental or commercial organizations to adapt quickly to a rapidly changing OE. Employing a PSC can keep military forces available to conduct traditional or specialized military missions.

5-23. Evolution of private sector military-like services by corporations can be a very influential factor in international and regional diplomacy. Although coalition operations may appear in need of services from PSC, governmental authorities and private citizen groups can be concerned on the level of PSC transparency and accountability when high-profile incidents occur that involve PSCs.

5-24. PSCs often hire large numbers of indigenous people from the local area. The vetting procedures may not screen out some undesirable elements of the population. A legitimate PSC may be susceptible to insider threats that infiltrate elements of the organization and subvert legitimate practices into forms of corruption. PSC security for the governing authority or commercial enterprises can be a conduit for corruption that disrupts support to a governing authority and/or coalition forces. The irregular OPFOR can cooperate with or coordinate corruption and extortion by local or regional warlords and criminal organizations. Private security subcontractors can be warlords, strongmen, commanders, and militia leaders who compete with the

Chapter 5

governing authority for power and authority. Providing "protection" services empowers PSC leaders with money and legitimacy. In some cases, PSCs can become private armies. Irregular OPFOR operations under the guise of a PSC can thrive in areas that appear to be devoid of government authority.

Organization

5-25. A PSC can be fairly small in scale or a large corporate organization. Figure 5-3 shows an example of possible functions in a PSC organization. The board of directors acts much like a headquarters staff of a military organization. Support services provide administrative, financial, and logistics support. Information and investigative services collect, analyze, and disseminate all types of information and intelligence, including economic intelligence. They also perform personnel security investigations and counterintelligence functions. Security services provide active and passive security measures that may be either rudimentary or use advanced technology. Guard services provide bodyguards, personal security details, stationary guard teams, and surveillance teams. Patrol services conduct patrols and tactical may conduct actions such as an ambush (to counter an intrusion). Security, guard and patrol service members receive weapons training.

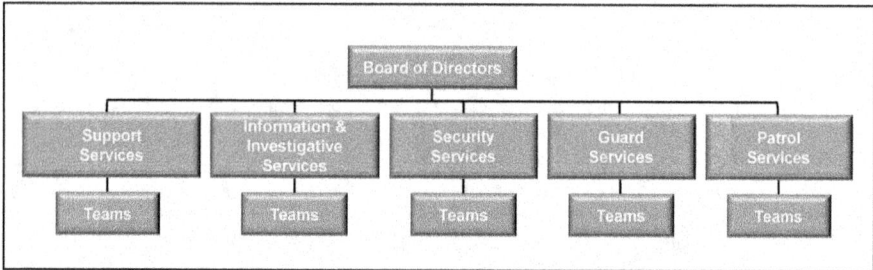

Figure 5-3. PSC organization by functions (example)

5-26. A PSC organization is tailored to serve its customer's needs. However, its organizational structure and level of capability are often directly related to a client's ability to pay. The capabilities of PSC employees vary from highly trained former military members to uneducated, poorly trained recruits. The level of competence and the sophistication equipment depends on what the client can afford.

5-27. The weapons and equipment mix is based on team specialization and role. Equipment might include—

- Listening and monitoring equipment.
- Cellular phones.
- Cameras.
- Facsimile machines.
- Computers.
- Motorcycles.
- Helicopters.
- All-terrain vehicles.
- Light armored vehicles.
- Assault rifles.
- Machineguns.
- Submachineguns.
- Antitank grenade launchers.
- Antitank disposable launchers.
- Silenced weapons.

LOCAL BUSINESS OWNERS AND EMPLOYEES

5-28. Local commercial enterprises often have authorization to own personal firearms to protect their workplace from theft or robbery. Authorization may be for either concealed or open carry. A business owner or employee with a weapon is not necessarily a threat to the governing authority or coalition forces. However, the mere presence of many weapons in public places adds to the uncertainty whether those forces should consider such armed noncombatants as friend or foe. The irregular OPFOR can use this type of uncertainty to slow the actions and reactions of the governing authority and/or coalition forces.

5-29. A local business may hire a PSC to protect its assets and interests. The irregular OPFOR or criminal organizations may coerce a businessperson into running a front company on their behalf.

CRIMINALS AND CRIMINAL ORGANIZATIONS

5-30. Criminals and the irregular OPFOR in an OE can be associated and/or affiliated with each other. For example, some insurgents and criminals can form temporary coalitions when it serves their mutual interests. Crime is a lucrative means to—
- Fund operations.
- Coerce and control key leaders and/or a relevant population.
- Erode governmental authority.

5-31. Some criminals may oppose the irregular OPFOR when its actions jeopardize criminal operations and profits. Criminal organizations may hire PSCs to provide additional security. Larger criminal organizations can employ paramilitary elements and tactics. Some may expand from traditional criminal activities into a pseudo-insurgency and establish de facto governance in areas, regions, and enclaves within an otherwise sovereign territory of the governing authority. (See chapter 4 for a more detailed discussion of criminal organizations and their possible roles in support of and/or cooperation with the irregular OPFOR.)

PRIVATE CITIZENS AND GROUPS

5-32. Local citizens often have authorization to own personal firearms for the personal security. Some may carry weapons as part of their occupation (such as hunters, security guards, or local police). Some are members of legitimate private clubs for shooting competitions or hunting. In some societies, carrying a weapon is simply part of the local culture. Many armed noncombatants openly carry weapons. This generally includes the individuals mentioned above, as well as members of public law enforcement agencies and various types of internal security forces. Other organizations and individuals may be authorized to carry concealed personal weapons.

5-33. Ad hoc local militia or neighborhood watch programs may provide some of the same functions as a PSC. However, they do so without some of the more sophisticated means and without the profit motive. These ad hoc local citizen organizations perform such actions on a volunteer basis to protect their families, homes, and businesses. (A local citizen organization that performs these functions on a contractual basis is a PSC.)

5-34. Such a large number of armed noncombatants in public places makes it harder to distinguish them from combatants. Members of the irregular OPFOR also try to blend in with the local populace when they are not conducting operations. The irregular OPFOR can use this type of uncertainty to slow the actions and reactions of the governing authority and/or coalition forces.

Note. In a broad sense, all individuals and groups not part of a formal, regulated military force are civilians. That includes insurgents, guerrillas, and criminals.

UNARMED NONCOMBATANTS

5-35. Other actors in an OE include unarmed noncombatants. These nonmilitary actors may be neutral or potential side-changers in a conflict involving the irregular OPFOR. Their choice to take sides depends on their perception of who is causing a grievance for them. It also depends on whether they think their interests are best served by supporting the governing authority. Given the right conditions, they may decide to purposely support hostilities against a governing authority that is the enemy of the irregular OPFOR. Even if they do not take up arms, such active support or participation moves them into the category of unarmed combatants. (See the section on Unarmed Combatants at the end of this chapter.)

5-36. Some of the more prominent types of unarmed noncombatants are—
- Media personnel.
- Nongovernmental organizations (NGOs).
- Transnational corporations and their employees.
- Private citizens and groups.

However, unarmed noncombatants may also include internally displaced persons, refugees, and transients. They can also include foreign government and diplomatic personnel present in the area of conflict.

MEDIA PERSONNEL

5-37. An area of conflict attracts a multitude of media personnel. This includes local, national, and international journalists, reporters, and associated support personnel. They may be independent actors or affiliated with a particular news organization.

Capabilities and Vulnerabilities

5-38. The media can be a credible source of current information for multiple actors in an OE. News cycles demand timely information and near-simultaneous reporting on current events. The irregular OPFOR recognizes the value of media coverage of significant incidents. This coverage can draw attention to irregular OPFOR successes or highlight failures or missteps by its enemies. The irregular OPFOR can exploit media coverage to attack the will and resolve of—
- Its enemies' regular military and internal security forces.
- The governing authority.
- A relevant population.

Media coverage of operations can dramatically affect international relations and strategic interaction.

5-39. The irregular OPFOR may closely observe media personnel. This surveillance can identify character flaws or weaknesses of personnel the irregular OPFOR can co-opt to the advantage of its INFOWAR activities. Although media personnel may seek to remain objective and report accurately, they can be coerced or persuaded to promote an irregular OPFOR perspective. The irregular OPFOR uses threats, extortion, and/or physical violence to minimize media coverage that is counter to its interests. Some media representatives who support irregular OPFOR motives may purposely distort information to support irregular OPFOR objectives.

5-40. The balancing effect of multiple reporting sources tends to reduce the impact of any one source with overt bias. The irregular OPFOR recognizes that democracies with freedom of the press and widespread access to media and other information systems can be less susceptible to INFOWAR. However, the international public and foreign governments are still susceptible to how the irregular OPFOR presents its agenda to a global audience.

Exploiting Media Access

5-41. The pervasive presence of the media provides access to information that might not otherwise be available to the irregular OPFOR. The irregular OPFOR can use the physical access allowed to media representatives to enhance its intelligence collection, information analysis, and consequent actions. Some members of the irregular OPFOR or individuals who support them may be able to pass themselves off as independent reporters or embed themselves in a media team under the guise of functional media expertise. Media credentials can be easily counterfeited. The embedding may take place with or without the knowledge of a sponsoring media organization. This may enable them to access plans and monitor operations of the governing authority or regular military forces with which the irregular OPFOR is in conflict.

Media Organization Structure

5-42. A media entity can be as small as an independent reporter or videographer operating as a free-lance entrepreneur or as large as a sophisticated media team with the credentials of a major news corporation or international sponsor. Well resourced reporters or media teams may have—
- Cellular telephones.
- Cameras.
- Video cameras.
- Tape recorders.
- Notebook or laptop computers.
- General office equipment.

They may have an all-terrain vehicle equipped with the latest video manipulation technology and satellite communication links.

5-43. Figure 5-4 shows an example of possible functions and individuals in a media team organization. While most media teams carry out most of these functions, they do not necessarily have all these specific coordinators and teams.

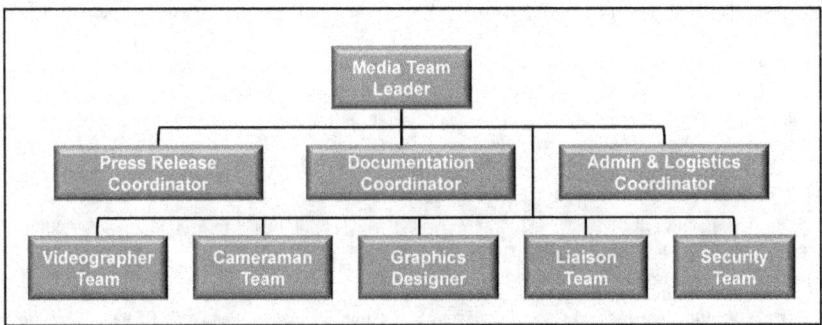

Figure 5-4. Media team organization (example)

Media Team Leader

5-44. The media team leader is responsible for the overall media coverage and news release program. This individual supervises the development of products that the media team presents. He promotes the value of its media resources and products and continually seeks to expand business investment in the team products.

Press Release Coordinator

5-45. The press release coordinator is the promotional "face" with the general relevant population and its influential leaders. He develops the persona of a trusted member to relevant population.

Documentation Coordinator

5-46. The documentation coordinator transforms raw audio or video coverage and produces a final package for release to the media-at-large via multiple means. Production coordination includes adding music, verbal announcements, graphics, and other sound and visual elements to enhance a compelling message. This coordinator is also responsible for crafting the long-term and recurring reinforcement of a particular message in conjunction with the press release coordinator.

Administration and Logistics Coordinator

5-47. The operational administration and logistics of the media team is the responsibility of the administration and logistics coordinator. Functions include—

Chapter 5

- Personnel management.
- Transportation.
- Lodging.
- Technical support.
- Repair services.
- Communications connectivity.
- Transmission security.

The administration and logistics coordinator coordinates with the security team in planning, occupying, and departing designated media coverage areas.

Videographer Team

5-48. The videographer team uses video camera equipment to record an assigned project. It then edits and assembles an audio-visual production. The documentation coordinator and videographer team review the product to ensure consistency and relevancy in support of an intended message. Each team member is responsible for the operation and maintenance of the equipment used in shooting and editing. A videographer may also be responsible for lighting and sound at a particular event. Depending on the assigned task, videographers can use diskette, videotape, and/or live broadcast capabilities.

Cameraman Team

5-49. The cameraman team is responsible for physically operating a camera and maintaining composition and camera angles throughout a given mission or task. Still-shot images are the norm for a cameraman. Duties are similar to that of a videographer.

Graphics Designer

5-50. The graphics designer has expertise in emphasizing visual impact to using graphics or other special effects such as animation. This individual also coordinates with media points of contact on how to best promote the news message and image of materials including electronic and other media venues. Much of the graphics design is focused on what effectively communicates with the targeted audience. This audience is usually influential individuals of a relevant population in an irregular OPFOR area of responsibility (AOR) as well as governing authority officials that are in conflict with the irregular OPFOR.

Liaison Team

5-51. The liaison team works closely with community members and leaders to ensure access to areas and information that supports the media mission or task. However, other teams may conduct their own liaison.

Security Team

5-52. The security team is responsible for the physical security and protection of media team members in the conduct of their duties. Security includes the actual mission sites, movement from and to the media team's operating base and/or logistics support areas.

Relationship to OPFOR Public Relations

5-53. Like media affairs, OPFOR public relations involve focused efforts to understand and engage key audiences of a relevant population. Their purpose is to create, strengthen, or preserve conditions favorable for the advancement of OPFOR interests, policies, and objectives. They accomplish this through the use of coordinated programs, plans, themes, messages, and products synchronized with all regular OPFOR and irregular OPFOR actions.

5-54. Public relations are part of the perception management element of INFOWAR. In an expanding INFOWAR campaign, the irregular OPFOR seeks partners within the specific OPFOR AOR and regional and/or global supporters. When external states provide overt and/or covert support, the irregular OPFOR provides appropriate public relations guidance on how to portray or hide such support.

5-55. A significant audience external to an irregular OPFOR AOR can be the diaspora of a relevant population. INFOWAR and public opinion are critical to obtaining diaspora support and keeping the struggle of guerrillas and/or insurgents in the spotlight of globalized media.

NONGOVERNMENTAL ORGANIZATIONS

5-56. A *nongovernmental organization* (NGO) is a private, self-governing, not-for-profit organization dedicated to alleviating human suffering; promoting education, health care, economic development, environmental protection, human rights; supporting conflict resolution; and/or encouraging establishment of democratic institutions and civil society (JP 3-08). NGOs are likely to be present in any OE.

Variety of Types

5-57. The global community of NGOs includes a wide variety of organizations that are independent, diverse, and flexible. They differ greatly in size, resources, capabilities, expertise, experience, and missions. An NGO may be local, national, or transnational. It may employ thousands of individuals or just a handful. It may have a large management structure or no formal structure at all. It may be a large organization with a huge budget and decades of global experience in developmental and humanitarian relief or a newly created small organization dedicated to a particular emergency or disaster. NGOs are involved in such diverse activities as education, technical projects, relief activities, refugee assistance, public policy, development programs, human rights, and conflict resolution.

5-58. Some NGOs represent members of a particular religious faith. In most of these cases, humanitarian aid originates in faith-based principles, but promoting a particular religion is not part of the aid provided. However, some missionary aid groups use humanitarian aid as a means of access for religious awareness and conversion of a relevant population.

5-59. Within a particular OE, there may be local NGOs as well as an international support presence of larger NGOs. International NGOs may use localized staffs in building capabilities in and for a region. Sometimes larger NGOs from outside a geographic region may not be openly visible while operating through these local organizations. This cooperation can provide resources unavailable at the local and regional levels during a crisis while incorporating localized capacity and knowledge of conditions, interpersonal relationships, and informal operations.

NGO Organizational Structure

5-60. A board of directors (sometimes called executive committee) is normally part of the large-scale decisionmaking process and working issues of the organization. The board serves as the trustee body of the NGO. Board members are valuable in that they extend the organization's resources into various communities and include personalities that are not specifically significant in daily operations but provide credibility to the organization as a whole. Many NGO boards have celebrities, former government officials, experts, academics, and community leaders with the intention of attaining recognition or publicity. During emergency appeals, NGOs will often send board members out to make public statements, write newspaper or journal editorials, make speeches, or give interviews to spark focus on the organization's work and needs in responding to the emergency.

5-61. NGO boards vary widely in style. Some are very active, often establishing close relationships with NGO officers and staff and involving themselves in planning processes and fundraising programs. Other boards are largely fundraising entities, lending their names to give credibility to an organization's fundraising practices. Boards can be primarily symbolic and fulfill the legal requirements of meeting a specific number of times each year and certifying financial obligations. The actual authority may reside in leaders who purposely remain discrete or unpublicized donors of an NGO.

Note. The irregular OPFOR may seek to co-opt board representatives and/or influential celebrities and donors to overtly or covertly support its goals and objectives.

5-62. Figure 5-5 shows an example of the possible functions of an NGO field office. A smaller, local NGO may not have all these functions.

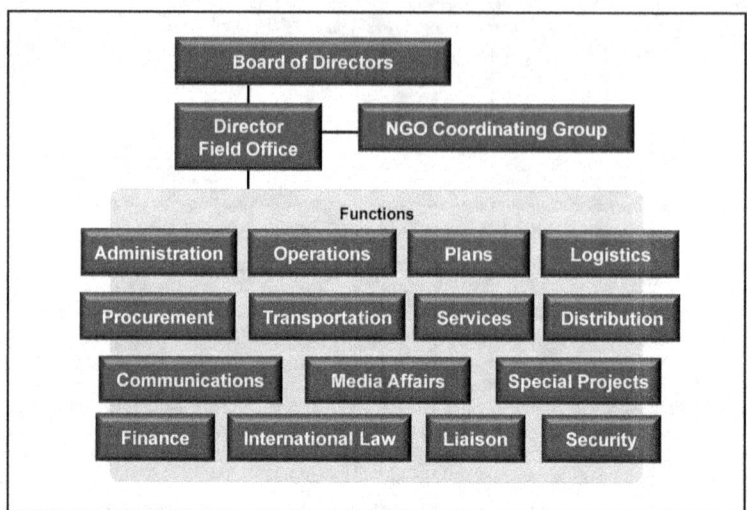

Figure 5-5. NGO field office functions (example)

5-63. Internal coordination, though not always formal, is the most common way civilian sectors organize in humanitarian emergencies. NGOs coordinate extensively within the local NGO community to manage information. They can act as advocacy bodies to international donors or local and regional governments. There is also external coordination between NGOs, regular military forces, governments, and donor agencies involved in a geographic region and/or international setting.

Capabilities and Vulnerabilities

5-64. Some internationally active NGOs may employ indigenous groups as local implementing partners. Extensive involvement, local contacts, and experience gained in various nations make NGOs valuable sources of information about local and regional affairs and civilian attitudes.

5-65. Many NGOs are outfitted with very little, if any, equipment for communications and personal security. They prefer instead to rely on the good will of the local populace for their safety. However, many larger NGOs have very capable communications capabilities. Some NGOs provide considerable support and assistance in the security area.

5-66. Communications provide some special challenges to NGOs. NGOs strive to provide assistance based on need rather than race, religion, political affiliation, or ability to pay. To avoid the perception of favoritism and operate in a neutral manner, NGOs normally do not use any encryption devices. They operate with the understanding that the local governing authority will likely be monitoring all their communications. They often have access to sensitive information that could assist political or military objectives of either side. This requires that information be recorded and transmitted in a way that reduces rather than incites conflict.

5-67. Modern communications equipment is essential for the NGO headquarters to remain in contact with field offices or mobile teams. However, such equipment might make NGO personnel a target of crime in areas where the technology is not commonly available. A satellite phone, which provides an NGO team global coverage, might represent a significant capability if acquired by the irregular OPFOR or criminals. The same might apply to a vehicle outfitted with a communication suite of sophisticated radios. In some areas, simply carrying a cellular phone or radio could be viewed as a threat by actors who would like to keep their activities concealed.

5-68. Security planning by an NGO is an ongoing process that continually reassesses threats. In some cases, NGOs operate in violent environments without developing a strong organizational culture of safety and security. Aid personnel often rely on the apparent immunity provided by a perceived humanitarian space and do not view themselves as party to the conflict. Various factors, including programs connecting an NGO and a governing authority, can cause specific targeting of NGO staff for kidnapping or murder by violent groups. NGOs often have formal safety and security systems to meet these threats. Some NGOs use professional security officers and/or PSCs with former military, police, or intelligence experience.

5-69. Reliable transportation is essential for NGO operations. However, this also poses the risk of accidents, assault, injury, and death for aid workers. The NGO must balance the need for a completely reliable vehicle against the risk of becoming a target of carjacking in areas of conflict or crime. NGOs may hire local drivers who have knowledge of which routes to use and how to negotiate check points. Such drivers may eventually gain trust as representatives of the organization. A driver from a group disenfranchised due to religious, ethnic, racial, or caste, may cause prejudice against the NGO. A driver might also be part of a criminal gang or the irregular OPFOR with the intent on stealing the vehicle or committing a crime against the NGO.

NGO Cooperation with a Governing Authority or Coalition Forces

5-70. The mission of an NGO is often one of a humanitarian nature and not one of assisting a military force in accomplishing its objectives. International guidelines on NGOs emphasize the principle of "humanitarian space" (humanitarianism, neutrality, and impartiality).

5-71. The preferred external coordination for NGOs is a United Nations (UN)-coordinated effort rather than a military-administered effort unless the UN is unable to do so. In cases where civil government is incapable or not present to coordinate NGO response, NGOs and militaries can interface with each other in operations centers administered by regular military forces.

5-72. Large NGOs often employ specialists or retired military personnel to assist in managing relations and activities that require interface with military entities. NGOs and military units have many incentives to coordinate and collaborate. NGOs often need many things from the military: logistics assistance, communications, intelligence, and protection. Military forces often find value in coordinating with NGOs for understanding ground-level activity, and to manage population movements, assistance projects, and general humanitarian activities. Both groups find that coordination is essential.

5-73. Military personnel often contact a NGO coordination group, which can—
- Provide contact information for liaison with member NGOs.
- Indicate what NGOs are doing in the region or area.
- Identify which NGOs are active in specific areas.

5-74. The UN often sets up humanitarian information centers (HICs) as central coordination points. NGOs are either required or voluntarily register with an HIC to keep updated on information and regional developments. A civil-military operations center or humanitarian operations center can be established for official coordinating between NGOs, the UN, governments, intergovernmental organizations, and the military.

5-75. NGOs and military forces can operate together, but challenges exist. Sometimes, in a region receiving NGO support and despite coordination efforts and extensive collaboration, NGOs can become frustrated with the apparent idle capacity that military commanders have in reserve for emergency or contingency actions. NGO managers may question why large numbers of military personnel, expertise, and supply stores remain unused while NGOs are short-staffed and significant support requirements exist for displaced persons and other elements of a relevant population. NGOs often want to ensure a distinct separation between military operations and NGO relief activities.

5-76. Coordinating and cooperating among NGOs and regular military forces does not necessarily mean sacrificing NGO impartiality or transparency. The benefits of working with military forces can enhance more effective and efficient NGO operations. Some factors for and against cooperation are—
- NGOs are generally capable organizations but may lack robust communication and logistics capacities.

- NGOs are civilian organizations that do not fall into a military-style command structure.
- NGOs prefer to keep their programs, activities, and image as independent as possible.
- NGOs can work effectively and efficiently during an emergency, often with a comparative advantage, in specific areas or regions.
- Routine and informal meetings between the NGO community, UN agencies, and the military can reduce much of collective tension and misunderstanding.

Motivations

5-77. NGOs are generally motivated by charity. They are trying to help the local population deal with manmade and natural disaster and disease, hunger, and poverty. NGOs are not affiliated with a government, but may be influenced by governing authority policies of their home nation.

5-78. NGO leaders and operators may have both stated and hidden interests and objectives. Some organizations and individual participants may have motivations that differ from the NGO's public mission statements. These motivations can be—
- Political.
- Economic.
- Ideological.
- Religious or theological.
- Social or cultural.
- Private and/or personal agenda.

The irregular OPFOR may play on these motivations to induce NGO personnel to act in ways that favor its own objectives.

Relation to Irregular OPFOR

5-79. Individuals participating in or in conjunction with an NGO may willingly support or be coerced to support the irregular OPFOR. The irregular OPFOR seeks to embed its supporters and/or covert operatives in NGOs operating its AOR. This could allow it to access plans and monitor operations of the governing authority or regular military forces with which it is in conflict.

5-80. Perceptions can easily be interpreted as reality by a relevant population when INFOWAR presents a compelling irregular OPFOR message and story of unresolved grievances. The irregular OPFOR can manipulate NGO actions to knowingly or unknowingly serve its goals and objectives.

5-81. The irregular OPFOR can use visible cooperation by an NGO with regular military or internal security forces of a governing authority to its advantage in INFOWAR. Media exploitation by the irregular OPFOR can depict NGO presence as biased in favor of a governing authority that is unjust and corrupt. Reducing NGO presence or causing NGOs to depart particular geographic regions may increase the civil turmoil the irregular OPFOR desires in its actions against the governing authority.

5-82. NGOs can be inundated with large numbers of civilians seeking medical care in refugee camps and NGO facilities. It may be difficult for an NGO to distinguish sick or wounded members of the irregular OPFOR from others requesting care.

5-83. NGOs provide essential services, comfort, and hope to those in need. Unfortunately, the irregular OPFOR may exploit the charitable sector to support its organizations and activities. This abuse can take many forms, including:
- Establishing front organizations or using charities to raise funds in support of irregular OPFOR organizations.
- Establishing or using charities to transfer funds, other resources, and operatives across geographical boundaries.
- Defrauding charities through branch offices or aid workers to divert funds to support irregular OPFOR organizations.

- Leveraging charitable funds, resources, and services to recruit members and foster support for irregular OPFOR organizations.

TRANSNATIONAL CORPORATIONS

5-84. A transnational corporation is a business that conducts commerce beyond national boundaries. Corporate activities may be regional or global. The term *transnational*, as opposed to *international*, accentuates the fact that these organizations and activities are not created or governed national governments. Instead, they arise under private enterprise for profit.

5-85. Transnational corporations may enter into partnerships with countries that are trying to increase their world economic position. Emerging states may invite such corporations to establish research and manufacturing facilities in their countries as a means of building infrastructure. The presence of these corporations can also enhance a country's security. However, the motivations of transnational corporation leaders are business oriented and not usually charitable. The corporations may try to influence regional affairs or assist their host country in actions that promote their own economic gain.

> *Note.* In pursuit of its own financial gain, a transnational corporation may provide overt or covert support to the irregular OPFOR. This may include support in obtaining weapons and equipment, fiscal funding, transshipment of materiel to irregular OPFOR sanctuaries, and/or placing economic pressure on governing authorities.

5-86. When external forces become involved in a particular country or region, they must take into account transnational corporations conducting business in the region. The presence of these outside business interests can put additional pressure on the intervening external forces to avoid collateral damage to civilian life and property.

> *Note.* Some transnational corporations may have their own armed security forces, or hire PSCs, to protect their business interests or perhaps also those of the host country and governing authority. Such security forces become *armed* noncombatants. However, if they participate in armed conflict on behalf of the host country, they become combatants.

5-87. A host country must also take into account transnational corporations conducting business within its borders or region, and their international connections. If the host country's actions adversely affect foreign commercial enterprises or regional security, conditions may encourage local, regional, or international support for irregular OPFOR goals and objectives.

PRIVATE CITIZENS AND GROUPS

5-88. The various types of civilians in an OE can include—
- Government officials.
- Business people.
- Farmers.
- Lawyers.
- Doctors.
- Clergy.
- Tradesmen.
- Shopkeepers.
- Other groups of the local relevant population.
- Transients.
- Internally displaced civilians.
- Refugees.

These individuals and activities include offices, bureaus, and agencies of a governing authority, intergovernmental organizations, NGOs, civil organizations, and independent or transnational corporations.

They provide a wide range of services, support, and other activities that can be susceptible to the influence of the irregular OPFOR.

> *Note.* The irregular OPFOR can attempt to embed operatives in low-level jobs in order to infiltrate government, commercial, and military activities. Such an insider threat can reap significant information and intelligence for the irregular OPFOR. It can also increase the restrictions placed on a relevant population and/or incite additional dissatisfaction with policies and practices of the governing authority.

5-89. Some government officials, such as police and emergency service personnel, have specialized equipment and standard uniforms. Other officials, such as mayors and town council members, wear clothing appropriate to the local customs. These officials are generally linked to a geographical area and/or functional responsibility and authority. Each actor has varied concerns and agendas. In some instances, the irregular OPFOR can masquerade as officials and act in a manner to discredit the governing authority.

5-90. Various political, religious, or social groups may be affiliated with the irregular OPFOR. Even if unarmed, this can affect their status as noncombatant actors. Criminal organizations or elements of a criminal organization can—

- Be unarmed noncombatants who violate civil law.
- Be armed noncombatants.
- Be involved in armed combatant acts.

> *Note.* The irregular OPFOR identifies weaknesses in the social structure and/or individuals of a relevant population and seeks to co-opt or manipulate particular individuals and/or groups to enhance its agenda. It may also acquire organizational capabilities, skill sets, and/or materiel in the population within which it operates. When possible, it gains this support voluntarily. However, it uses extortion and violence to attain its objectives if required.

5-91. Support from the civilian population is a key to irregular OPFOR operations. The irregular OPFOR recruits its manpower, gains intelligence, and receives safe haven and materiel support from the relevant population. Even if it also has external support, the irregular OPFOR cannot exist for an extended period without significant support from the civilian population.

UNARMED COMBATANTS

5-92. The local populace contains various types of unarmed nonmilitary personnel who, given the right conditions, may decide to purposely support hostilities against the enemy of the irregular OPFOR. Such active support or participation may take many forms, not all of which involve possessing weapons.

5-93. In an insurgent organization or guerrilla unit, unarmed personnel might conduct recruiting, financing, intelligence-gathering, supply-brokering, transportation, courier, or INFOWAR functions (including videographers and camera operators). Technicians and workers who fabricate improvised explosive devices might not be armed. The same is true for people who provide sanctuary for combatants.

5-94. Unarmed religious, political, tribal, or cultural leaders might participate in or actively support the irregular OPFOR. Unarmed media or medical personnel may become affiliated with a military or paramilitary organization. Even unarmed individuals who are coerced into performing or supporting hostile actions and those who do so unwittingly can in some cases be categorized as combatants.

> *Note.* From the viewpoint of some governing authorities, any armed or unarmed person who engages in hostilities, and/or purposely and materially supports hostilities against the governing authority or its partners is a combatant.

5-95. Individuals who perform money-laundering or operate front companies for large criminal organizations might not be armed. Individual criminals or small gangs might be affiliated with a paramilitary organization and perform support functions that do not involve weapons.

Chapter 6

Terrorism

Terrorism is a tactic. This chapter presents an overview of conditions that are a composite of real-world capabilities and limitations that may be present in a complex operational environment that includes terrorism. Acts of terrorism demonstrate an intention to cause significant psychological and/or physical effects on a relevant population through the use or threat of violence. Terrorism strategies are typically a long-term commitment to degrade the resilience of an enemy in order to obtain concessions from an enemy with whom terrorists are in conflict. International conventions and/or law of war protocols on armed conflict are often not a constraint on terrorists. Whether acts of terrorism are deliberate, apparently random, and/or purposely haphazard, the physical, symbolic, and/or psychological effects can diminish the confidence of a relevant population for its key leaders and governing institutions. Social and political pressure, internal and/or external to a relevant population and governing authority, is frequently exploited by terrorists with near real-time media coverage in the global information environment. The local, regional, international, and/or transnational attention on acts of terrorism by state and/or non-state actors can often isolate an enemy from its relevant population and foster support of organizations, units, or individuals who feel compelled to use terror to achieve their objectives. The themes and messages promoted by terrorists can accent anxiety, demoralize the resolve of a relevant population and its leaders, and eventually defeat an enemy.

TERRORISM IN COMPLEX OPERATIONAL ENVIRONMENTS

6-1. *Terrorism* can be defined as the use of violence or threat of violence to instill fear and coerce governments or societies. Often motivated by philosophical or other ideological beliefs, objectives are typically political in nature. The pursuit of goals and conduct labeled as terrorism by some actors in complex operational environments (OE) can be considered fully justifiable by other actors. The spectrum of actors in an OE can range political, public, and/or commercial institutions, other institutions appearing legitimate but disguising an illicit agenda, and/or organizations and individuals who openly declare intent to use terror as a matter of policy and practice. Irregular forces typically use terrorism (see figure 6-1).

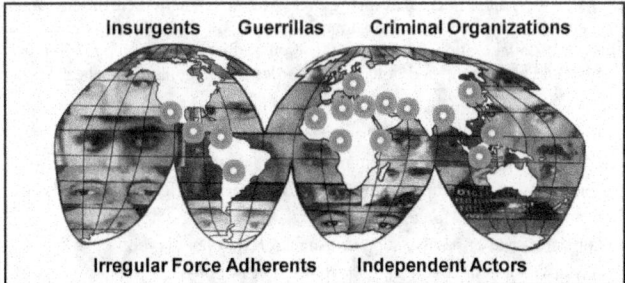

Figure 6-1. Terrorism actors in complex operational environments

6-2. The geographical perspectives of local, national, regional, international, and/or transnational terrorist acts can indicate why terrorism is used but a more effective measure of understanding terrorism must relate directly to the actors involved in a particular OE. Ultimately, terrorism is about people and the psychological effectiveness of convincing or compelling a relevant population to act in a desired manner.

6-3. Multiple actors in an OE can use terrorism to enhance the conditions that favor their agenda and overarching purpose. Terrorism in irregular actions favors indirect and asymmetric approaches. However, terrorism can be employed within a full range of military and other explicit or subtle capabilities in order to erode an enemy's power, influence, and will. Irregular forces are generally categorized as armed individuals or groups who are not members of the regular armed forces, police, or other internal security forces of a governing authority. Similarly, a hybrid threat can be visualized as the diverse and dynamic combination of regular armed forces, irregular forces such as insurgent organizations and/or guerrilla units, and/or affiliated and associated criminal elements all unified to achieve mutually benefitting effects. Other irregular forces can include mercenaries, corrupt officials of a governing authority, compromised commercial enterprises, public entities, active or covert supporters, and/or coerced and passive citizens of a populace.

6-4. Independent actors using terrorism can also exist in an OE distinct and separate from the agendas from other irregular forces. Adherents may have no direct connection with an irregular force but apply terror in support of an irregular force's agenda. Threats can be as small as one individual or expand in size from a small and discrete band, cell, or unit to the noticeable presence as a large group of rogue actors that are affiliated or united in purpose.

6-5. Irregular forces adapt their capabilities in an agile and flexible manner to achieve organizational objectives. Terrorism is a means of conducting violent conflict that typically provides significant psychological impact on an intended target. Acts of terror may be actually intended for an audience other than the victims of an attack. Conducting terrorism typically emerges from a deliberate decisionmaking process. Terrorists usually compare and contrast advantages and disadvantages to evaluate probable cost-benefits of a particular act or acts of terror. When an irregular force terrorist self-justifies the legitimacy of using terrorism against an enemy, the terrorism often affects enemy combatants and noncombatants in a relevant population. The analysis and decision to use terror is often a simple cost-efficient and effective results-oriented means to pursue violent conflict in order to achieve a purpose.

Note. For the purpose of this TC, the term *enemy* refers to an enemy of the irregular force.

MOTIVATIONS

6-6. Motivations often include a wide range of reasons that an individual, cell, unit, or other type of irregular force organization may feel compelled to use terrorism. Encouragement from outside an organization to use terrorism can be part of the conditions and circumstances that influence an irregular force to act. Support to use terrorism may be from regular military forces and/or special-purpose forces (SPF) from a state or states. Internal security forces and/or law enforcement organizations that have been infiltrated by an irregular force can also support irregular force actions within a local or regional area of responsibility. The collaboration among organizations, units, cells, and/or individuals may be based on coercion, contractual agreement, and/or temporary or long-term common goals and objectives.

6-7. Motivations for initiating and/or continuing acts of terrorism can include—
- Spotlight attention on unresolved grievances with an enemy.
- Disrupt an enemy's ability to continue actions against an irregular force.
- Champion causes of a suppressed and/or disenfranchised segment of a relevant population.
- Demonstrate irregular force capabilities.
- Obtain active and/or passive support from a relevant population.
- Receive overt and/or covert support from a state or non-state actor.
- Deter continued enemy military operations in a particular geographic area.
- Dissuade enemy governmental influence over a relevant population.
- Develop acceptance and legitimacy of an irregular force agenda and programs.
- Cause an enemy to overreact to acts of terror and correspondingly alienate a relevant population.

- Defeat enemy military and/or internal security forces and/or its governing institutions.
- Achieve irregular force objectives.

6-8. Irregular force objectives promote solutions to grievances in the context of a particular relevant population. An irregular force may prefer to use indirect approaches such as subterfuge, deception, and non-lethal action to achieve its objectives, but is committed to violent action when necessary to compel an adversary, enemy, and/or other opposing form of governance to submit to the irregular force demands. In most cases, irregular force operations include politically oriented plans of action. Some irregular force organizations, such as affiliated criminal gangs, exist for their own commercial profit and power and are not interested in improving aspects such as quality of life and civil security of a relevant population that they affect.

6-9. Transforming a grievance into a concept and plan for action develops typically along a pattern that evolves from generalized ideas to tactical options and a heightened sense of needing resolution to a grievance. Continuous information and intelligence collection refines options for the likelihood of success and a decision to act. Motivation provides a momentum of commitment to actually conduct the terrorism. A way to visualize this human dimension sequence is a terrorism planning cycle.

TERRORISM PLANNING AND ACTION CYCLE

6-10. A generic sequence and timing of irregular force terrorism depends on organizational capabilities and limitations, operational constraints, and the level of commitment of an irregular force actor or organization. To effectively understand this commitment, knowing the underlying motivation is fundamental to appreciating the resolve to plan and act. The irregular force sets conditions to optimize its awareness, training, and mission readiness to achieve objectives that counter enemy forces. When advantageous to irregular force operations, coordination and cooperation can combine the capabilities of conventional military, paramilitary, criminal activities, and/or terrorism.

6-11. Tactics, techniques, and procedures include creating conditions of instability in a particular OE, alienating the population from the governing authority of the region, and improving the irregular force influence on a designated populace and key leaders in that relevant population. In complex conditions, an irregular force may be able to employ a range of organizational options from small loosely affiliated cells to global networks in order to promote psychological effect and mission success. Such networks can be local, regional, international or transnational affiliations; host simple or sophisticated media affairs programs; as well as acquire covert and/or overt financial, political, military, or social support.

6-12. A terrorism planning cycle is actually a continuum. Irregular forces plan, prepare, act, and apply experience and skill in order to achieve objectives. The concept of a spiral effect may be an effective way to visualize and understand a planning cycle of terrorism. Even with periodic setbacks in acts of terror capabilities and execution, the resolve of terrorists to a compelling agenda is often progressive, adaptive, and long-term in order to achieve objectives.

BROAD TARGET SELECTION

6-13. Irregular force terrorism operations are typically prepared to minimize risk and achieve the highest probability of success by avoiding an enemy's strengths and concentrating attack on an enemy's weaknesses. Emphasis is often placed on maximizing irregular force security and terrorism effects. Security measures usually include planning and operating with small numbers of irregular force members to more effectively compartment knowledge of a pending terrorism mission.

6-14. Collection against potential targets may continue for years before an operation is decided upon. Detailed planning is a norm but can be deliberately shortened when an opportunity arises. While some targets may be vulnerable enough with shorter periods of observation, the information gathering and analysis for intelligence will be intense. Operations planned or underway may be altered, delayed, or cancelled due to changes at the target or in local conditions. Tactical missions conducted by and/or for larger irregular forces complement operational objectives and strategic goals of the irregular force. The psychological impact on a targeted population is the overarching objective of any terrorist operation.

6-15. There is no universal model for planning, but irregular forces use their experience and expertise to effectively apply traditional principles for plans and operations. Irregular forces often exchange expertise in particular skill sets such as recruitment, media affairs, and training on various forms of direct action in terrorism. Tactical methods and analysis of successful missions are often shared via the Internet and websites hosted by an irregular force. Adaptability, innovation, improvisation, and risk assessment are key components of plans and actions toward mission success (see figure 6-2).

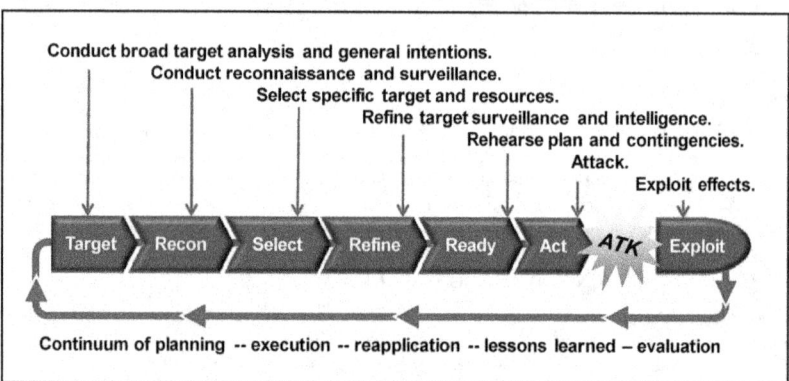

Figure 6-2. Terrorism planning and action cycle

6-16. Tactical and operational planning can be analyzed according to common requirements. A plans and operation cycle provides a baseline in assessing particular organizational requirements to conduct an act of terror. The significant differences among irregular forces often center on factors of intent and capability and the organizational, key leader, and individual commitment to a philosophical, ideological, or otherwise practical motivation and compulsion.

6-17. Irregular forces often pride themselves on being a learning organization. Combined with motivation and a compelling agenda, irregular forces gather information and intelligence, analyze their own and enemy strengths and weaknesses, determine enemy patterns, trends, and emerging actions susceptible to attack, and identify key vulnerabilities in an enemy's systems, functions, and actions.

6-18. A consideration in selecting potential targets involves actions to counter enemy governing authority as supporting force programs that demonstrate security, safety, and assurance to the relevant population that the irregular force needs to influence. Tactical operations that might counter an enemy through irregular force subversive or direct actions can include—

- Secure a critical segment of a relevant population.
- Declare the control and/or sovereignty of geographic areas that support the irregular force.
- Disrupt military and/or internal security forces of an enemy.
- Diminish the credibility of civil police and law enforcement activities.
- Degrade a public sense of safety and stability from the enemy governing authority.
- Discredit the political agenda of the enemy governing authority.
- Demonize key leader actions of the enemy governing authority and civil administration.
- Provide basic social services not being adequately provided by the enemy governing authority.

INITIAL INTELLIGENCE GATHERING AND SURVEILLANCE

6-19. Targets displaying significant vulnerabilities may receive additional attention and priority of effort for intelligence gathering and surveillance. This potential for successfully attack establishes a requirement to gather additional information on a target's patterns over time. This phase may be very short or can span months and even years. The type of surveillance employed depends on the type of target and its location. Elements of information typically gathered include—

- Assessing the practices, procedures, and/or routines of an organization and/or facility. Items of interest include scheduled product deliveries, work shift changes, identification procedures and other observable security routines. Examples are as simple as recording when regularly scheduled supply deliveries arrive or commodity pickup occurs, or where key leaders park personal vehicles.
- Observing the physical layout and individual activities at residences and business offices of key leaders. Actions include knowing how the power grid services the location and critical points of failure if electrical power is to be interrupted during an attack.
- Monitoring transportation routes of travel for targeted individuals for common routes, choke points, and limited visibility areas along routes that may be conducive for an attack. Personal patterns can include routes among a personal residence, temporary lodging, commercial site, gym, and/or school or university. Attention at facilities may include access and exit points, types of vehicles allowed on the facility property, and/or physical security barriers that must be bypassed or breached to enter a location.
- Probing security measures to determine the complexity of security at a target and/or reaction time of security response units. Other items of interest can include any hardening of structures, barriers, or sensors; personnel, package, and vehicle screening procedures; and the type and frequency of emergency reaction drills at a facility.

SPECIFIC TARGET SELECTION

6-20. Selection of a target for actual tactical planning typically considers some of the following issues:
- Does success affect a larger audience than the immediate victim(s)?
- Will the target attract immediate high profile media attention beyond the immediate region?
- Does attack success emphasize the desired grievance to the appropriate audience that can resolve the grievance?
- Is the attack effect consistent with overarching objectives of the irregular force?
- Does the target and mission success provide an advantage to the irregular force by demonstrating its organization's capabilities?
- What are the near-, mid- and long-term costs versus benefits of conducting the operation?

6-21. A decision to proceed requires continued intelligence collection against the selected target. Targets that do not receive immediate primary consideration may still be collected on for future opportunities.

PRE-ATTACK SURVEILLANCE AND PLANNING

6-22. Members of the unit or cell that will conduct the attack begin to appear during this phase. Trained intelligence and surveillance personnel or members supportive of the irregular force may be organized to prepare logistics and/or locations for the operation. Operatives gather detailed information on the target's current patterns in recent days to weeks. The irregular force assesses and confirms information gathered from previous surveillance and reconnaissance activities.

6-23. The type of surveillance employed depends on the target's activities. Current information is used to—
- Conduct security studies.
- Conduct detailed preparatory plans and operations.
- Recruit specialized operatives and/or active supporters.
- Procure a base of operations in the target area such as safe houses and caches.
- Design and test escape routes.
- Decide on types of weapons and other means of attack.

REHEARSAL

6-24. Rehearsals improve the probability of success, confirm planning assumptions, and assess contingencies. Rehearsals test security reactions to particular attack profiles in primary and alternate plans of attack. Irregular forces use their own operatives and/or unsuspecting people to test target reactions.

6-25. Typical rehearsals include—
- Equipment and weapons operational checks.
- Communications and signals to be used in the mission.
- Skills performance of all and/or particular specialists.
- Final preparatory checks.
- Pre-operations inspection drills.
- Deployment sequence of movements and maneuver into the target area.
- Actions near and/or on the objective.
- Primary and alternate escape routes.
- Initial safe haven, hide sites, and/or rally point actions.
- Transfer plans from initial to subsequent safe havens or hide sites.

6-26. Confirmation checks in the target area can include—
- Target information gathered to date.
- Target patterns of activities.
- Physical layout of target area for changes in routes and/or manmade features.
- Time-distance factors from the assault position to the attack point.
- Security force presence during varied states of alert.
- Reaction response timing by security forces to a demonstration, feint, and/or threat.
- Ability to preposition and/or retrieve equipment or vehicles near the objective.
- Ease of blocking and/or restricting an escape route at critical choke points.

ATTACK AND ACTIONS ON THE OBJECTIVE

6-27. The irregular force executes its tactical plan but remains flexible to changing conditions and adapts accordingly. The plans and rehearsals have considered primary and contingency actions that possess the advantages of initiative and deception. Actions provide for—
- Use of Surprise.
- Choice of time, place, and conditions of attack.
- Employment of diversions and supplemental attacks.
- Support of security and related positions and/or forces to neutralize target reaction forces and security measures.

6-28. Simultaneous actions may include an assault element, security element, and support element. Some missions may require a breach element. Actions on the objective will sequence through several main tasks:
- Isolate the objective site.
- Gain access to the individual, individuals, and/or asset.
- Control the target site.
- Seize and/or destroy the individual, individuals, and/or asset at the objective.
- Achieve the mission task.

ESCAPE AND EXPLOITATION

6-29. Escape plans are well rehearsed and executed. Rapid withdrawal and dispersal from the target site can involve multiple withdrawal routes and temporary safe houses. Even in the case of a suicide attack, a handler or observer-recorder requires a plan to evade identification and capture. Similar expectation to evade capture occurs with an attack by fire element or support by fire element.

6-30. Media exploitation to a global audience will usually include video coverage, sometimes with audio commentary by observers with a videographer. Mass casualties or major disruption of economic and social services typically gain prime international media coverage. Extending the duration of an act of terrorism may promote awareness of an irregular force agenda and assist the irregular force in winning the "battle of the narrative" with a relevant population.

6-31. The irregular force attack must be actively publicized to achieve an intended effect of exploiting a successful attack. Media outlets influenced or coerced to support an irregular force agenda with prepared public statements are examples of preparation to effectively exploit irregular force operations. Release of warnings and/or announcements are timed to take advantage of media cycles for selected target audiences at the local, regional, and global levels of information warfare.

6-32. Unsuccessful operations are often disavowed by the irregular force when possible. The perception that an irregular force has failed can damage its prestige and indicate vulnerability.

6-33. In addition to the negative impact on the enemy, successful attacks can bring favorable attention, notoriety, and encourage support such as funding and recruiting to the irregular force. The proof that an attacker can attack and evade may sway potential recruits to join and/or convince recruits that have been coerced to approach assigned tasks with an expectation of success and survival.

6-34. The general concept of planning and action by irregular forces is adaptable. The seven concept phases are generally descriptive of tactics and techniques but are not prescriptive. Operations retain a flexible expectation based on evolving conditions. Knowing the underlying motivations for terrorism is fundamental to appreciating the resolve of irregular forces to plan and act.

6-35. When advantageous to irregular force operations, capabilities can include combinations of regular military forces, paramilitary units, insurgent organizations and guerrilla units, and criminal networks. Any and/or all of these groups can use terrorism. Regardless of an irregular force network as local, regional, international or transnational in capability, media exploitation using simple or sophisticated systems is critical to create the desired influence within a targeted population.

TERRORISM ACTIONS

6-36. Irregular forces apply terrorism for offensive and defensive purposes. Common tactical principles of armed conflict and INFOWAR apply capabilities with the intent to cause psychological anxiety and fear in an enemy in addition to physical effects. Adding aspects of surprise and deception to acts of terrorism often weakens resolve of a targeted relevant population. Nonetheless, each situation in an OE can present tactical variations and techniques to conduct a successful irregular force mission.

6-37. In applying a definition of tactics as an ordered arrangement and maneuver of forces in relation to an enemy to achieve a mission objective, an irregular force has a wide range of technique options in traditional or irregular conflict. Techniques describe methods used to conduct a tactic in order to accomplish required functions or tasks. Techniques evolve through the analysis of an intended mission and observations and/or lessons learned from successes and failures of previous missions. Procedures are standardized steps, performed deliberately and consistently, that prescribe how to perform specific tactical functions and tasks. The "how-to" of understanding terrorism is a composite of knowing TTP and the motivations that provoke and compel their use by irregular forces.

6-38. Incidents of terrorism can range the rogue action of a lone individual or the sanctioned activities of large organizations acting on the overt or covert policies of a state. Terrorism can also be sponsored or conducted by non-state actors. The following descriptions focus primarily on individual and small unit, cell, or group actions experienced as a sudden, violent engagement among friendly and enemy forces. These tactical engagements usually orient on offensive actions and terrorism but may also require irregular forces to transition temporarily to defensive forms of conflict.

6-39. The irregular force uses a flexible array of terrorism means and materiel to accomplish assigned missions. The irregular force makes decisions under conditions of uncertainty but continually seeks to minimize its risk and identify vulnerabilities in an enemy that can be attacked. Deception and surprise compound the effects of massing an attack against a point of weakness in order to achieve physical effects and create anxiety or fear. Understanding and applying these debilitating effects of coercion, anxiety, and/or fear is central to the intended physical and psychological effects of terrorism.

6-40. The irregular force is adaptive and learns from tactical success and failure and will adjust techniques to particular conditions in order to achieve an objective. TTP examples underscore the fact that the science of tactics is only as effective as the leadership, training, and experience of an irregular force unit, cell, or related organization. Elements such as demonstrated capabilities, weapon systems, location, restrictions and constraints, logistic support, and time-distance and weather and terrain factors are important to planning and conducting an irregular force action. However, the essential aspect of executing an irregular force act is the motivation and commitment of an irregular force in the individual and collective execution of tasks to achieve a mission objective.

DEFENSIVE ACTIONS

6-41. Irregular force tactics can augment defensive actions with terrorism in order to—
- Defeat an enemy attack.
- Gain time.
- Economize irregular force capabilities.
- Develop conditions favorable for subsequent offensive operations.

6-42. Other objectives for conducting defensive actions related to terrorism include denying access to an area, causing extensive commitment of enemy forces and materiel, and/or fixing enemy forces for a specific time. Since acts of terrorism are typically sudden, violent acts similar to an assault, ambush, or raid, defensive actions often involve the delay or disruption of enemy response to the act of terrorism. An irregular force may attempt to deny access to designated terrain or a resource for a specific time, limit the freedom of maneuver to an enemy pursuit, or channel enemy forces into a killing zone for a subsequent assault or ambush. Defensive actions remain closely linked to offensive actions and terrorism.

OFFENSIVE ACTIONS

6-43. An irregular force plans and acts decisively with deception and surprise as tactical enablers for terrorism. Creating overwhelming combat power against a specified objective requires an irregular force to have keen situational awareness and understanding of its AOR.

6-44. The shock effect of terrorism depends heavily on effective reconnaissance and surveillance in information gathering and intelligence analysis. The resulting awareness and understanding improves the ability to combine irregular force effects at a time and place for the optimum application of surprise and deception with available resources for an expectation of mission success.

Attack Threat on Land

6-45. Attacks using terrorism are often considered most likely during the normal conduct of lifestyles in familial, social, or commercial settings. Attacks often occur in the context of a local setting. Political perspectives can affect each of these daily domains and the willingness of people to be an active or passive supporter to an irregular force agenda.

6-46. An irregular force evaluates the situational context of its operational environment in order to determine the most effective ways to advance its objectives. The variables of a particular setting and the capabilities that exist for improvised or innovative actions by an irregular force can include physical domains beyond conflict on land or its sub-surface. Two other domains can be maritime and aerial environments in which to conduct irregular conflict.

Maritime Attack Threat

6-47. A maritime attack has the potential for significant disruption and/or damage on the economy of a governing authority and other international actors with which the irregular force is in conflict. An attack at a major port facility or a choke point on a main maritime route would cause significant regional disruption and negative impacts on a global economy.

6-48. If a weapon of mass destruction (WMD) was detonated at any of these locations, the immediate damage would be great to shipping and port facilities. However, a potentially even greater damage would be the economic repercussions on regional and global markets and commerce.

Aerial Attack Threat

6-49. An aerial terrorism threat can use multiple means to attack. Traditional forms include hijacking aircraft and hostage-taking, sabotage of aircraft and facilities, and/or using a manned civil or military aircraft as a suicide bomb against a designated target.

6-50. Another variation is use of an unmanned aerial vehicle (UAV) with an explosive payload to attack a target. Techniques can range from small hobby craft model airplanes with small ad hoc explosive packages to the acquisition and use of sophisticated cruise-type missiles with significantly larger warhead capabilities.

TACTICS, TECHNIQUES, AND PROCEDURES

6-51. Irregular forces use a wide array of tactics and techniques to apply terrorism. The TTP are intended to be flexible and adaptive approaches. Surprise, secrecy, and indirect methods of attack are fundamental to acts of terror. The tactical options are as broad and diverse as the resolve of the irregular force leader to improvise and/or innovate with available resources.

6-52. The irregular force leader uses tactics, forces, and weapon systems tailored to a particular mission. Operations are planned for a specific target and effect and use reconnaissance and surveillance to plan, counter, and overmatch an enemy. If changes or unexpected conditions render success unlikely, the irregular force leader may cancel or postpone an operation and return at a more opportune time. He may choose a different target and continue his planning and attack process.

6-53. Extensive use of the Internet encourages exchange of practical information, training, and observations among terrorists. Training material, training videos, and on-line dialogues develop and sustain initiatives and encouragement among irregular forces. Shared information and intelligence continue to improve irregular force techniques as field experiences demonstrate degrees of effectiveness in employing acts of terror.

6-54. Tactics, techniques, and procedures typical of irregular force actions and terrorism include—
- Threat-Hoax.
- Arson.
- Sabotage.
- Bombing.
- Hijack-Seizure.
- Kidnapping.
- Hostage-taking.
- Raid or ambush.
- Assassination.
- Weapons of mass destruction (WMD).

THREAT OR HOAX

6-55. Irregular forces can use threats to coerce or preclude actions by a targeted individual or population. Threats and hoaxes can degrade the effectiveness of preventive or countermeasures when a targeted individual or population loses situational awareness of an actual terrorist threat or disperses finite assets against many possible threats. At the less lethal end of the conflict spectrum, hoaxes can simply be methods to annoy and/or degrade the alertness of security forces and keep the population constantly agitated. Bomb threats, leaving suspicious items in public places, and other ploys consume time, resources, and effort from an enemy and its security operations and contribute to general uncertainty and anxiety.

6-56. Extortion is an example of a threat that can obtain money, materiel, information, or support by force. For example, numerous reports from a region could indicate that warlords and militias are in contractual agreements with a governing authority to protect government and coalition supply convoys. However, these same warlords and/or militia in many cases are corrupt and extort large sums of money from the regional trucking contractors. Corrupt government officials or an activity that require bribes to retain business permits or easy route access further complicates actual or alleged threats and extortion. Another example of extortion can occur in the opium trade of a region with "taxes" levied on farmers, traffickers, and competing trafficking networks.

6-57. Intimidation is another form of extortion. An irregular force individual, intelligence cell or specialized team can intimidate people to obtain information on a target location or to provide resources. Death threats against an individual or his family may cause the individual to provide information or resources to a cell, unit, or organization with which he has no interest or allegiance. Intimidation can also be used as a means to convince individuals not to take an action. For example, security personnel may be intimidated to not implement required security measures during a specified period of time at a critical infrastructure facility.

6-58. The effects of coercing individuals can be significant while appearing irrational to some people. Nonetheless, irregular forces have successfully used these simple techniques to coerce individuals to participate in an attack that could include an individual suicide or vehicular borne suicide mission.

ARSON

6-59. Arson is most often used for symbolic attacks and economic effects. Examples of attack by irregular forces can include—
- Assaulting and setting fire to a temporary staging area for fuel tanker trucks.
- Burning transportation infrastructure in order to disrupt the flow of logistics along a main supply route and a resulting requirement for reallocation of security forces to reinforce route protection.
- Damaging communications nodes to disrupt cable or digital relay transmissions.

6-60. A number of arson attacks by small groups of armed extremists can hamper other ongoing governing authority operations and/or regional law enforcement responsibilities. Arson can be combined, directly or indirectly, with other forms of terror. An arson attack on residential areas or public facilities could injure or kill noncombatants and incite civil unrest.

SABOTAGE

6-61. Sabotage is the planned destruction of the enemy's equipment or infrastructure. The purpose of sabotage is to inflict both psychological and physical damage. Sabotage can be an incident creating a large number of casualties or cause a severe disruption of services for a relevant population.

6-62. Destroying or disrupting key services or facilities displays irregular force capabilities on the public consciousness and either increases a target population's frustration with the ineffectiveness of its governing authority or may inspire others in the population to resist along with the irregular force. Oil pipelines, water purification plants, sewage treatment facilities, air traffic control hubs, and/or medical treatment or research facilities are examples of potential sabotage targets. Sabotage and terrorism can also be demonstrated as a combination of techniques such as bombing, arson, cyber, or use of industrial materials and contaminates.

BOMBING

6-63. Bombs are commonly used by irregular forces to conduct acts of terrorism. Bombs can be highly destructive and easily tailored to a mission, do not require the operator to be present, and have a significant physical and psychological impact (see figure 6-3).

6-64. The type of terrorism can employ varied explosive device —
- Military-grade type-classified mines.
- Military-grade or civilian engineer/construction explosives.
- Commercial-industrial products adapted to create explosive components and/or devices as improvised explosive devices (IED).

6-65. Car bombs, commonly referred to as vehicle borne improvised explosive devices (VBIED), are used regularly in terrorism. Suicide vehicle borne improvised explosive devices (SVBIED) are a norm of some terrorists. In a contemporary tally of attack methods used by irregular forces, only armed attacks surpass the use of bombing.

Bombing: Insurgent Affiliate (Example)

An insurgent organization affiliate operating within a sovereign state selected an urban compound that housed coalition forces as the target for a vehicle borne explosive device (VBIED) attack. The governing authority had been successful in suppressing any public demonstrations by the insurgent organization as coalition forces arrived in the state to assist in stabilizing the expanding insurgency of the region. A small cell of six insurgents conducted reconnaissance and detailed surveillance of the compound and identified that breaching an entry into the compound was not likely to succeed. The plan evolved to detonate a large VBIED close to a perimeter wall near high-rise apartments inside the compound. The fabricated bomb was sure to cause significant death and injuries to the multinational military members and civilian contractors operating from the compound.

Insurgents placed several thousand pounds of explosive inside the bulk fuel tank of a commercial tractor-trailer truck. Training and other logistical support was provided by the insurgent organization in a neighboring state of the region. An insider threat identified that the governing authority would not allow any expansion of stand-off distance along a particular section of compound wall. Insurgent observers confirmed that physical security checks by local police were infrequent and the area next to the wall was easy to access.

At an early morning hour with visibility limited due to a heavy rain, three insurgents waited near the compound wall in a sedan. Other insurgents drove the tractor-trailer fuel truck up to the compound wall, set the timing device on the bomb, and quickly departed the area with the other insurgents in the sedan. An insurgent observer remained to observe security force and first responder reactions to the bombing.

Guards inside the compound noticed the suspicious activity but were unable to warn many of the residents in the apartments before the VBIED detonated. Structure damage was significant to several apartment buildings. The bombing caused 35 deaths and over 100 other people were injured in the attack.

The public outcry from several coalition states caused the compound to be evacuated within days. The insurgent organization claimed a major victory in forcing the removal of coalition forces from the area as the insurgency increased its active and passive support of the relevant population.

Figure 6-3. Bombing: insurgent affiliate (example)

6-66. A technique used often by terrorists is an initial bomb detonation followed by secondary bomb detonations as first responders or other people arrive at the attack site. In recent years, bombs and in particular IEDs increased in lethality and adaptation of techniques used by terrorists. Some IEDs are bulky devices often made from artillery shells and detonated with simple electric triggers using garage door openers or doorbells. However, irregular forces also produce and use smaller lethal devices that can be planted quickly and can be detonated from longer distances with more sophisticated devices.

6-67. Another IED innovation is to use a device called an explosively formed projectile (EFP). The penetrating principle is common to shaped-charge munitions in order to outmatch armor protection. Although some technical skill and machining is required to obtain an optimum effect, a simple EFP is a section of pipe filled with explosives and capped by a shaped copper disk. When the explosive detonates, the EFP liner folds into a slug-like shape to penetrate armor plating of a vehicle. Emplacement of the EFP is factored to hit a most likely weak point for penetration. The angle of an EFP attack can be from below a target, along a side of a target, and can even be from above or at an extended height from a road or trail surface to attack wheeled or tracked vehicles.

6-68. A prevalent suicide tactic involves an individual wearing or carrying an explosive device to a target and then detonating the bomb or driving an explosive-laden vehicle to or into a target and then detonating the bomb. Suicide attacks differ in concept and execution from other high risk operations. In other high-risk missions, mission success does not require that the participant die. The plan allows for possible escape or survival of the terrorist or terrorists.

6-69. Some terrorists have used people who are unknowingly part of a suicide attack. An example is an individual associated with a terrorist cell who believes he is only a courier of information and materiel but is unknowingly transporting an IED in a vehicle that is command- detonated by an observer-handler against a selected target.

6-70. Another way of describing a suicide bomber is a highly effective precision-guided munition. Psychological impact increases when confronted by a person who plans to intentionally commit suicide in order to kill other people. Although a suicide bomber can be a lone terrorist working independently, the use of suicide terrorism as a tactic is normally the result of a conscious decision on the part of the leaders of an irregular force to engage in this form of attack.

HIJACK-SEIZURE

6-71. Hijacking involves the forceful commandeering of a means of transportation such as an airplane, ship, train, or bus. Purposes for hijacking and terrorism can include—
- Hostage-taking activities.
- Obtaining a means of escape from a tactical area of responsibility.
- Obtaining a means to conduct a suicide attack with a weapon of mass destruction.

6-72. While hijacking of aircraft for hostage taking has declined in frequency in recent times, the use of hijacked aircraft for escape or as a weapon continues to be a dangerous threat. The attacks on the World Trade Center and the Pentagon in September 2001 are tragic examples of hijacking commercial jet planes that are used for terrorism in a deliberate and multiple suicide attacks.

6-73. The use of hijacked vehicles for destructive devices is not restricted to aircraft. Trucks carrying explosive or flammable materials have been seized for use as weapon delivery devices. The possibility of such a technique could also apply to commercial transport ships carrying petroleum products such as liquefied natural gas (LNG). The damage or environmental contamination caused by an explosion of such carriers or other explosive commodities would be devastating. Accidental explosions of this nature are evidence of the catastrophic infrastructure damage that would be caused to ports and people.

6-74. Seizure of a critical element of infrastructure, similar to hostage taking, can be a physical site such as a facility of critical importance to a relevant population or a cyber node that disrupts and/or prevents use of selected cyber functions. The terrorism threat of disruption or destruction of seized infrastructure can be a bargaining issue for an irregular force or could be the intentional seizure and use of infrastructure to contaminate and/or damage a geographic area.

KIDNAPPING

6-75. Kidnapping is usually an action taken against a prominent individual for a specific reason. However, kidnapping can be an indiscriminate act to cause anxiety in a relevant population. The most common reasons for kidnapping include—
- Ransom.
- Barter for release of a fellow irregular force member.
- Desire to publicize a demand or issue with terrorism.

6-76. Success of kidnapping often relies on correctly assessing the impact to the governing authority in maintaining their prestige with a relevant population. A contrast is the possible reactions of the governing authority to demonstrate their resolve. Detailed planning considers contingencies of how to release the kidnap victim and/or how often to move the kidnap victim among safe houses. The irregular force typically states specific demands with a timeline for a response. Kidnapping can also be used as a means of financing the organization. Ransom from seized individuals or groups is often a significant income for irregular forces in several regions of the world. Irregular forces are often willing to hold a victim for significant time in order to have its demands satisfied (see figure 6-4).

Kidnapping: Insurgent-Criminal Affiliation (Example)

A criminal organization was contracted by an insurgent leader to kidnap a prominent member of the regional governing authority. This appointed government official was a leading advocate to disrupt the cross-border smuggling operations of the criminal organization and had recently sponsored the arrest and detention of several criminal leaders. Members of the criminal organization observed the daily routine of the official and determined that seizing him at his residence provided the best opportunity for success. No personal security existed once the driver and guard delivered the official to his apartment. The criminals posed as city water engineers inspecting a reported water leak in the apartment complex.

When the official allowed the engineers to enter his apartment, they gagged and bound him before placing him in a work chest on rollers. The official's spouse and children were gagged, blindfolded, and tied to a room radiator but were otherwise physically unharmed. The kidnap victim was casually rolled out of the apartment complex to a waiting van and brought to a safe house in a nearby town. The official was delivered to an insurgent cell and brought to a separate safe house.

The insurgent organization sent photographs of the official and a manifesto to media outlets and on the Internet that unless the criminal leaders were released from detention by the governing authority, the official would be killed. Negotiations continued for two weeks before the governing authority agreed to release the criminal leaders in exchange for the official. Although the official was released unharmed, the kidnapping embarrassed the governing authority in its apparent inability to curtail the criminal organization in its illicit international smuggling operations. Concurrently, the insurgency gained more active and passive support from the relevant population in the region's urban areas and sections along the inter-border smuggling routes.

Figure 6-4. Kidnapping: insurgent-criminal affiliation (example)

6-77. Some kidnapping operations are actually assassinations with killing the victim as a planned outcome. Actions include obtaining intermediate concessions and publicity during the negotiation process that the irregular force would not receive from a simple assassination. Irregular forces may distribute videotapes and/or photographs to the media and post similar exploitation means on Internet websites.

HOSTAGE-TAKING

6-78. Hostage taking is typically an overt seizure of people to—
- Gain publicity for a grievance.
- Seek political attention and/or concessions.
- Obtain safe passage for irregular forces contained by enemy forces during a mission.
- Demand release of irregular force prisoners held by an enemy governing authority.
- Ransom.

6-79. Unlike kidnapping where a prominent individual is normally taken and moved to an unknown location, the hostages can be individuals not well known in the enemy's society. Hostage situations are frequently risky for irregular forces especially when conducted in enemy territory. This location exposes the irregular forces to hostile military or police operations and carries the possibility of mission failure and capture. Therefore, hostages may be held in a neutral or friendly area rather than in enemy territory. Irregular forces will sometimes take often hostages with the intent to kill them after they believe they have fully exploited the immediate tactical advantage and/or media coverage from the situation. However, some hostage situations are conducted to cause a direct confrontation with enemy forces and the resulting global media attention while negotiations are in progress (see figure 6-5).

Hostage-taking: Guerrilla Unit (Example)

A guerrilla unit was directed by its higher insurgent organization to seize hostages and demand several political concessions from the governing authority with whom the insurgency was in conflict. The raid by approximately 15 guerrillas approached the opening day ceremonies of a middle school in a medium size industrial city. Guerrillas jumped from trucks firing weapons into the air as they entered the large school courtyard and quickly forced students, parents, and other relatives into the main school building. Local police had been deceived earlier by a hoax and were in a different part of the city. Government response to the hostage-taking was slow and ineffective.

Guerrillas barricaded the school building and after separating young male adults from the main group of hostages, murdered the 20 male adults and dumped their bodies out of a second floor window into a main street. Internal security police had arrived at the scene along with regular military forces but were unable to prevent the murder and mayhem.

International news media arrived on the scene based on insurgent announcements posted to the Internet. The governing authority's response to the terrorism received public criticism by the local citizenry and mass media as confused and sluggish. The guerrillas allowed selected negotiators and news representatives inside the school to record the tension and suffering of the hostages. Guerrillas stated their willingness to fight to the death if attacked or if their demands were not satisfied.

The stalemate ended abruptly when a guerrilla improvised explosive device detonated under suspicious circumstances. Both guerrillas and government forces started firing at each other and many hostages were killed or wounded in the firefight. Beyond the tragic loss of life and injuries, the governing authority was embarrassed in front of a global audience. All guerrillas died in final room-by-room assaults by government forces inside the school but the insurgent leader achieved his objective of spotlighting the plight of the relevant population in the region being suppressed by an unjust governing authority.

Figure 6-5. Hostage-taking: guerrilla unit (example)

RAID OR AMBUSH

6-80. A raid or ambush is similar in concept and can be used to accent anxiety in an enemy with its sudden and violent actions. A raid permits control of the target for the execution of some other action. The kidnapping or assassination of a target may require a raid to gain access and control a particular person or area for a brief period of time.

6-81. Some raids may not expect irregular force members to survive. Other raids may use raiders that volunteer to attack with the expectation of committing suicide as integral to the assault and/or to fight until killed in order cause the maximum amount of destruction and mayhem (see figure 6-6).

Terrorism

Raid: Multiple Insurgent Suicide Cell Assaults (Example)

An insurgent organization conducted nearly simultaneous assaults with a cell of two and three-person teams at multiple locations in a major metropolitan city. The choice of target areas ranged business centers, famous entertainment sites, mass transit hubs, and cultural landmarks for the purpose of indiscriminate killing as well as focused hostage-taking and murder of citizens and foreign visitors based on their nationality or religious faith. Terrorism was to exploit psychological trauma on a major urban population and undermine the confidence of citizens on the ability of their governing authority to protect its people and foreigners, as well as spotlight the insurgent organization's ideological extremism and agenda to global attention.

A ten-person direct action cell was thoroughly trained and indoctrinated in the safe haven of a neighboring state. None of the terrorists in the cell expected to survive the raid. The insurgent organization prepared for more than one year with refined surveillance and reconnaissance. The terrorist cell used a fishing trawler to infiltrate past coastal security patrols and arrive at the city's harbor area. The cell unloaded their weapons, explosives, and equipment and split into five two-person teams. Dressed as normal citizens with travel bags, they moved to their separate objectives by walking, taxi cab, or a small inflatable boat for travel further into the city's port.

Confirming via cellular telephone that all assault teams were in position, the terrorists used semiautomatic rifle fire and hand grenades to assault people in a rail station, hospital, urban streets, café, hotels, and religious cultural center. Improvised explosive devices (IEDs) had been hidden in taxis and were timed to explode well after the simultaneous attacks were underway. These taxi explosions caused additional confusion and mayhem in the first hours of the incident. After attacking a café, one team quickly joined a team assaulting an internationally famous hotel complex. Random killing continued as they started fires in the hotel and seized hostages. Hostages at a cultural center were tortured and murdered. Another team killed or wounded people randomly in a train station and a hospital. As these terrorists attempted to link up with other insurgent cell members they were intercepted and killed by internal security forces in a city street.

Several IEDs were emplaced with timers or tripwires at various sites in or near a hotel. The terrorists barricaded in the hotel were in regular communication with senior insurgent leaders in a neighboring state via satellite telephone, personal digital assistant (PDA) devices, and cellular telephones. The insurgent leaders and cell handlers provided ideological encouragement and tactical advice to the insurgents as the hostage crisis continued for several days with live-feed media coverage to a global audience.

Regular military forces and special purpose forces eventually killed nine terrorists and captured one terrorist but the insurgents had already achieved their information warfare objectives. The murder or injury of hundreds of noncombatants and days of mayhem in a metropolis embarrassed the governing authority in front of international partners and its citizens and demonstrated the insurgent's commitment to its extremist agenda.

Figure 6-6. Raid: multiple insurgent suicide cell assaults (example)

6-82. An ambush is another form of surprise attack characterized by terror, violent execution, and speed of action. The intended objective may be to cause mass casualties, assassinate an individual, or disrupt enemy security and stability operations. Explosives, such as bombs and directional mines, are a common weapon used in ambushes as well as antitank rockets, automatic weapons, and/or other small arms fires.

Assassination

6-83. An assassination is a deliberate action to kill specific individuals. Targets for assassination are often individuals that have notoriety in a relevant population such as political leaders, well-known citizens, and/or public spokespersons. This form of terrorism can also target individuals with no public or governing authority importance and appear indiscriminate. This type of targeting for assassination can intimidate and cause anxiety and fear and coerce the passive support of a relevant population to an irregular force and preclude the active support to a governing authority with whom the irregular force is in conflict. Extensive target surveillance and reconnaissance of engagement areas is critical to planning and conducting an assassination. Although many factors play into the decision, the target's vulnerabilities often determine the method of assassination.

6-84. Many targets of assassination are symbolic and are intended to have significant negative psychological impact on the enemy and the relevant population that it controls. For example, assassinating an enemy senior government official, a successful corporate leader, or a prominent cleric can demonstrate the inability of an enemy governing authority to protect its own people. Assassinating local representatives and/or leaders of a society can contribute to civil disorder while demoralizing members of a local governing authority and/or law enforcement organizations.

Technique Enablers

6-85. Techniques that take advantage of restrictions and constraints on an enemy operating within a relevant noncombatant population are a norm of irregular force actions. Limitations on an enemy can include social traditions, moral codes, and/or legal policies of a governing authority and its forces. Irregular forces can remain discrete and secretive within a population or can choose to be open in their operations when operating in or from a safe haven. In either case, irregular forces seek to create an environment of authority and relative protection in and from which to operate. Examples of these enablers include dispersion of the irregular force organization within a relevant population and/or the use of noncombatants of the population as human shields when confronting an enemy.

6-86. Actions by irregular forces demonstrate a tempo and/or pace based on the capabilities of the irregular force as it attempts to gradually or dramatically impact on its enemy. Attacks preferably employ surprise and deception with irregular force units, cells, or other organizations and also plan for the infiltration of insider threats to enemy activities and infrastructure. Several offensive action techniques that can easily compound a sense of anxiety, fear, and/or terror are use of suicide attacks, cyber attacks, and attacks with a weapon of mass destruction.

Disperse

6-87. Dispersion in complex rural and urban environments can degrade situational awareness and complicate enemy intelligence and targeting efforts. Urban areas offer excellent cover and concealment from enemy ground forces and airpower because building interiors and subterranean areas are hidden from airborne observation, and vertical obstructions hinder the line of sight to ground targets. Irregular force individuals, cells, units, or other organizations are often decentralized and purposely dispersed in locations. Similar cover and concealment in rural areas may be available in villages or natural terrain such as caves, tunnels, and/or every restrictive physical terrain.

6-88. Safe houses or sites facilitate irregular force ability to discreetly transit from one location to another by providing a place to rest, acquire resources, plan, prepare, and stage for acts. Active supporters in a relevant population and/or criminal organizations can facilitate such networks and trafficking from location to location in a local community and/or across regional and/or international borders.

Shield

6-89. Irregular forces can deliberately use noncombatants as human shields that limit enemy forces in using their weapons systems when noncombatants may be injured or killed. Noncombatants are sometimes prevented from evacuating likely engagement areas to ensure that irregular forces have human shields present in their immediate vicinity. These groupings can conceal movements and/or be a means of escape for an

irregular force after executing an attack. Some irregular forces purposely use the elderly, women, and/or children as human shields. Activities to create selective areas of human shields can include—
- Orchestrating work strikes.
- Fomenting mass rallies.
- Coordinating peaceful-appearing demonstrations.
- Coercing civilians to gather with and around an irregular force action, security, and/or support element.

6-90. Types of withdrawal actions or repositioning through human shields causes terror for the noncombatants involved and may allow an irregular force to regain a tactical initiative and renew offensive actions. Asymmetric techniques take advantage of typical restrictions on enemy rules of engagement and often reduce enemy capabilities to apply their full suite of weapon systems against an irregular force. Some defensive tactics are to—
- Disperse within a relevant population of noncombatants.
- Use noncombatants as a human shield during armed conflict with an enemy.
- Exploit positioning in close proximity to infrastructure such as hospitals, schools, or places of religious worship.
- Conduct INFOWAR manipulation of actions when enemy forces cause noncombatant casualties in combating the irregular force.

Women and Children Combatants

6-91. Women and children cover a broad range of irregular force capabilities. Women and children can be involved in the direct actions of an assault, raid, or ambush and/or be critical to the security and/or logistics support to an irregular force.

6-92. Media reporting often spotlights the contributions of women in irregular force operations. In some instances irregular forces may encourage their front organizations to promote the visible presence of women and children in their public awareness efforts. In other instances, irregular forces will co-opt women and children to be suicide bombers in sensational incidents that receive much attention in media affairs.

6-93. Less obvious but often just as dangerous are other forms of direct combative and support activities. Women and children can serve a valuable role as couriers when cultural custom or civil restrictions in a region allow them easier ability to transit public areas without the level of detailed personal searches that a man would normally experience at checkpoints of a governing authority and/or its internal security forces and law enforcement organizations.

6-94. From other support perspectives, women can be key operators in the logistical systems of irregular forces. Personnel operations can involve the recruitment of new members with a maternal image and be an example of how a woman can obtain status within a social-cultural setting. Recruitment of women and children can range from selective invitation of individuals perceived to be receptive to an irregular force agenda to overt coercion and forceful levy for manning irregular forces in a local community. Alcoholic beverages and/or drugs can be used at times to dull the ability of children to resist assisting irregular force activities. Child fighters may be indoctrinated or coerced to conduct specific tasks and can be forced to act in irregular force operations with little or no training.

6-95. However, training of particular skills such a bomb making or other functional capabilities can support the transition of recruits to be fully integrated members in an irregular force. Women and men may be responsible for establishing and maintaining safe houses for irregular forces that may require a periodic sanctuary on short notice. Women and men may also be the overt presence in a facility used for maintaining kidnapped people, cached supplies, or medical treatment sites. Similarly, fund raising and the informal connections within a community can use women and children as a significant symbol of family value and need. When public expression of an irregular force agenda is allowable in a locale or region with recognition as a legitimate political party, women can be a spotlight for media affairs coordination and publicity. In some cases, women are elected political spokespersons for an irregular force agenda.

6-96. Women and children can be fighters with a weapon, couriers of money or munitions, shields to male fighters who engage an enemy with small arms fire, and/or other forms of support operations for an irregular force. Women and children in irregular conflict are a significant combat multiplier for an irregular force.

Safe Haven

6-97. Safe havens are typically located in ungoverned or physical areas where irregular forces are able to organize, plan, raise funds, communicate, recruit, train, transit, and operate in relative security because of inadequate governance capacity by a governing authority with whom an irregular force is in conflict. In some cases, safe havens are located in semi-autonomous areas within a sovereign state while in other cases an enclave across international borders. Distant locations from an irregular force AOR may be more acceptable to sustaining irregular force operations.

6-98. A safe haven is often used to train regional and/or foreign recruits and integrate them into the tactical and/or logistical support operations of an irregular force. An extension of some safe havens is use of cyber connectivity and the ability to mask or distort Internet irregular force cyber points of origin. Recruitment, training, and encouragement can be particularly effective with recurring communications and encouragement to irregular force members, affiliates, and/or adherents.

Insider Threat

6-99. An insider threat and intent to conduct acts of terror involve the infiltration by an irregular force member or members into an area and/or facility. The insider threat is a practical and psychological perspective that consumes significant enemy resources to counter such invasive actions.

6-100. The "insider" is an individual who is authorized access to open or secure systems of an enemy or supporting organization that can include physical facilities and infrastructure as well as information systems. In the context of irregular forces and terrorism, an insider can be a member of an irregular force and/or an adherent of an irregular force agenda. In either instance in context of command and control and/or information system operations and infrastructure, the insider threat can—

- Disrupt general and/or specific operations of an enemy.
- Deny enemy use of particular systems and data.
- Corrupt or destroy selected information.
- Cause malicious damage or destruction to a facility and/or other infrastructure.

6-101. An insider threat mission can be planned as a one-time breach of enemy security measures or be planned as a recurring, long-term covert investment to gather information and intelligence. An insider threat can be a willing participant with an irregular force or be coerced into supporting actions that support irregular forces. In addition to tasks on gathering specific information and/or intelligence on enemy operations and procedures, a complementary task may be to identify weaknesses in a particular enemy capability or function.

6-102. Irregular force accomplices may already exist within an objective area and be able to assist the entry and acceptance of additional irregular force members as an insider threat. Uniforms, security badges and/or identification cards, and/or purposeful lapses in physical security measures are examples of insider accomplice support. Once an insider threat is established in an assigned objective area, it may be directed to perform a number of tasks or be focused with a specific function. Tasks can include—

- Remain discrete and report on current operations and personal observations in the objective area.
- Plan for acts of sabotage on designated systems and/or equipment.
- Attempt to subvert other actors in the objective area to support the irregular force.
- Conduct sabotage as critical opportunities arise.
- Support the actions of other insider threats of the irregular force.
- Commit, on order, violent acts on a designated target.
- Conduct a suicide attack on a designated target.

6-103. An insider threat or an irregular force observer in the objective area can sometimes capitalize on witnessing an incident that identifies individuals with a personal and/or ideological grievance and who has already been vetted for security clearance into an enemy's organization and/or facility area. This type of individual may be gradually co-opted by an irregular force observer and recruited as an insider threat. On other occasions, a personal grievance may incite an individual to act violently even though there is no direct connection to an irregular force. The INFOWAR campaign of the irregular force will manipulate such incidents to support its agenda and often claim responsibility for the violent and/or terrorism actions.

Information Warfare

6-104. Irregular force information warfare can disrupt popular support for a governing authority and its allies and/or coalition partners. Irregular forces can spread rumors or misinformation as a means to offset the official information from a governing authority and its military and/or internal security forces.

6-105. Irregular forces can use the Internet to disseminate their message as quickly as events occur. An immediate press release from a Website offers irregular force control of initial message content and media reporting. Irregular forces manipulate images and create special effects or deception that enhances the anxiety and/or fear of terrorism. Audiovisual coverage of irregular force successes are incorporated in irregular force recruitment efforts and sustain morale. Multimedia sites can display manufactured evidence of governing authority atrocities and war crimes to turn domestic and international opinion against the governing authority. Irregular forces use sympathetic media to reinforce their INFOWAR campaign (see Appendix A).

Suicide Attack

6-106. Suicide attack can be the act of an individual or a tactic planned and conducted by an irregular force to spotlight a grievance with an intentional self-destruction and incidence of mass casualties, death, and mayhem. Mass media attention is often focused on an organization's grievance when such a sensational action occurs. As a tactic, suicide attack seeks to degrade the resolve of a relevant population and obtain concessions from an enemy to an irregular force agenda. Organizational aims and ideological support to commit suicide in attacking an enemy is often complicated in motivations but is typically an action used when other more conventional means of combating an enemy are overmatched and produce nil results. A tactic of suicide attack can also be evaluated for an operational intent and/or strategy to obtain support for an irregular force agenda from state and non-state actors who might be otherwise unwilling to get involved in a social or political conflict of an irregular force.

6-107. In whatever means a suicide attacker uses, an irregular force member conducting such an act of terrorism has the ability to adjust actions until the moment of attack. This adaptive behavior makes a suicide attacker particularly dangerous. When the weapon is a suicide bomb, the moment to detonate a suicide vest or a vehicle borne improvised explosive device can be adjusted given conditions as they exist at a time and place of a planned attack. An attacker can estimate the degree of mayhem that will be caused by the explosion and change who and where to assault. In other instances, a handler of the suicide attack can command-detonate a bomb on a willing and/or unknowing individual carrying a bomb. The attack can be a deliberate assault on a specified target or can be intentionally an indiscriminate act to kill and/or main combatants and noncombatants. Suicide attack is primarily a psychological assault on an enemy and its supporters.

Cyber Attack

6-108. A cyber attack can be obvious or purposely masked to prevent an enemy from knowing a system has been or is about to be attacked. Cyberspace use can be a key enabler to commit espionage, subversion, and/or sabotage. From a terrorism perspective, cyber attack can create anxiety and fear in a targeted population or group of key leaders. An intrusion into a cyber system can be embedded with a time delay feature and/or be programmed to execute an action on order of a specific signal. Whether a cyber intrusion is from an external source or has been emplaced by an insider threat, the reasons for conducting terrorism with a cyber attack on an enemy can include—

- Sabotage cyber networks.
- Extort concession from an enemy.
- Damage enemy or supporting activities and infrastructure.
- Endanger people within a relevant population as part of an irregular force objective.
- Support other threat capabilities with a greater probability of success in mission conduct.

- Identify vulnerabilities in cyber systems and facilities.
- Evaluate enemy TTP for the ability to limit or prevent particular cyber attacks.

6-109. Irregular forces can employ a broad range of technology to support cyber attacks. Cyber technology is readily available from commercial or illicit markets in order to conduct terrorism. The ability to exploit advanced or low-cost technology and integrate with basic TTP is limited only by the fiscal means and application skills of a terrorist. TTP can use a range of capabilities simultaneously to gather critical information for planning and conducting a future while a cyber attack is in progress. TTP of a very simple nature include—

- Exploiting social media to collect information through false representation
- Gathering open source information from Internet or intranet electronic postings.
- Exploiting people through techniques such as phishing, spearphishing, or whaling expeditions.
- Infiltrating malware to disrupt, damage, or destroy cyber systems.
- Exfiltrating critical information from cyber systems that enables sabotage and other types of attack.

6-110. A model for planning and conducting cyber attack can be very sophisticated in attempt to breach layers of detection and protection. Other models can be quite simple in concept. A cyber attack model can include—

- Reconnaissance to identify and select potential targets.
- Weaponization of a computer operating system or software application.
- Delivery of malware by remote or physical access to a targeted computer.
- Exploitation by triggering a malware code in a targeted computer.
- Installation that creates an access point on a victimized computer and allows unauthorized entry and exit on a victimized computer and network.
- Command and control to send instructions to targeted computers previously installed with payload malware and provides the means to conduct a cyber attack.
- Actions to accomplish the intended cyber attack objectives.

Weapons of Mass Destruction

6-111. Listing a category of weapons of mass destruction (WMD) acknowledges a broad range of capabilities that specific terrorist organizations, units, cells, and/or groups would like to acquire. WMD for an irregular force can be as simple in concept as several tons of improvised explosive from fertilizer and chemicals to acquisition of a sophisticated nuclear device from a rogue state and/or contract technicians. Fundamental information on using materials to construct types of WMD is available on the open-source Internet. Once skills are acquired and precursor or selected material and equipment are obtained by terrorists, the possibility exists to threaten use and/or actually employ WMD for catastrophic results.

6-112. The types of WMD are categorized generally as chemical, biological, radiological, or nuclear (CBRN) materials. Sometimes WMD is described as nuclear, biological, and chemical weaponry. However, this is a limiting viewpoint on what can be used as a WMD.

6-113. Materials that are weaponized as WMD can include toxic industrial chemicals (TIC) and toxic industrial materials (TIM), biological agents, and/or radiological materiel. Low-yield and high-yield common explosives can also be configured to cause results considered a WMD. Although difficult to acquire by irregular forces, a nuclear demolition and/or bomb-like device is by its nature a weapon of mass destruction. Any of these weaponized materials can be a significant irregular force and/or terrorism capability.

6-114. WMD can be employed through numerous types of delivery systems. Once a dangerous biological agent is acquired, a simple means of employing the WMD can be letters contaminated with a biological agent being sent through a postal service to cause death and/or severe injury to those individuals who are contaminated. Additional impacts of such an attack are contamination and/or the potential contamination of facilities that the letters transited during the delivery process. Extensive decontamination requirements would disrupt or temporarily halt private and or commercial mail systems. The psychological impact on a relevant larger population would accent anxiety, fear, and frustration of how to prevent other similar attacks with a WMD (see figure 6-7).

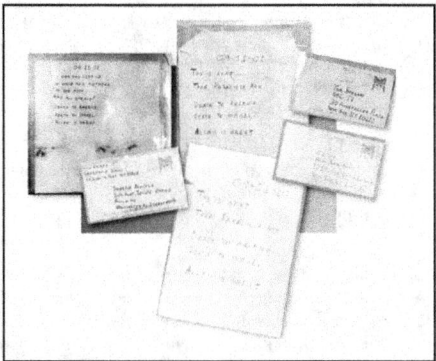

Figure 6-7. Anthrax-laced letters: death-mayhem in targeted population (example)

6-115. The expertise to acquire and use WMD remains a significant terrorist threat based on some irregular forces intent to—
- Acquire materials to construct or develop WMD.
- Threaten to attack with WMD.
- Attack combatants and noncombatants indiscriminately with WMD.
- Create mass injury, death, mayhem, and damage with WMD.

TERRORISM TRENDS AND EMERGENT VECTORS

6-116. Trends of terrorism can be considered emergent variations and adaptation to age-old truths of conflict when at least two forces are unequally matched in military power. Terrorism is often indicative of a weaker force seeking vulnerabilities in a more powerful force and attacking those weaknesses with an expectation of persistent conflict. No accurate prediction exists on the character, location, or duration of any specific future terrorism; however, irregular force will continue to improvise, innovate, and adapt TTP to achieve success through acts of terror.

6-117. Situational accelerators can spotlight probable vectors of terrorism adaptation by irregular forces in near-term years. Uncertainties include the pace and/or tempo an irregular force can demonstrate in its tactical or operational reach, effects an irregular force can display throughout the strategic depth of an enemy, and how successful the "battle of the narrative" will be in influencing a targeted population. The resilience of motives and ideological commitment of an irregular force and its leaders remain primary strengths in an irregular force campaign of terror. Terrorism trends and emergent vectors that irregular forces consider include—
- Improved situational awareness and understanding of the enemy.
- Radicalized philosophy and/or ideology as a compelling motivation.
- Adaptive organizational tactics and techniques.
- Versatile organizational affiliations.
- Expanded transnational affiliates.
- Emergent independent adherents.
- Devastating full spectrum weaponry.
- Targeted economic disruption.
- Globalized media affairs.

IMPROVED AWARENESS AND UNDERSTANDING

6-118. Understanding situational context approaches the issue of irregular emergent vectors along an avenue of action used by irregulars, paramilitary, guerrillas, terrorists, insurgents, or criminals. Intent has a premise of adapting constantly to optimize knowledge, training, logistical support, and readiness to conduct irregular operations. Terrorism will be used when this type of action accomplishes the desired psychological and physical effects. Irregular forces can be patient. Irregular forces plan while waiting for critical opportunities to strike.

6-119. Irregular force vectors consider the nature of irregular forces and the capabilities and limitations of specific irregular forces in an evolving contemporary operational environment. As the regions of the world advance in technological areas, expand the mobility opportunities of people, and exploit the Internet and other media, extremists concurrently fuel grievances and alienate segments of populations to foster support for their agendas.

6-120. Irregular forces assess and plan for conducting operations in an area of responsibility and also consider how to attack forces and/or states that may exist in regions that are distant from a primary area of land, air, and/or maritime conflict. Cooperation and collaboration among extremist affiliates and/or adherents will plan for and attempt to attacks the homelands of those states that may be supporting a governing authority with whom the irregular force is in conflict. Irregular forces also assess and plan for attacking intermediate staging locations that would disrupt the flow or prevent significant support from these states.

RADICALIZED MOTIVATIONS

6-121. Core grievances are real or perceived issues of segments in a targeted population. The importance of the core grievances or even their existence can change over time. Irregular forces can manipulate core grievances to create conditions for developing the willing support within a population. The perception of foreign exploitation or a governing authority that appears to be excessively influenced by foreigners can be a core grievance. For example, if foreign businesses dominate critical portions of the local economy, some of the population may feel that they or their country are being exploited by outsiders. A foreign military presence or military treaty may offend national sentiment as well. The mere presence or specific actions of foreigners may offend local religious or cultural sensibilities. A state or regional governance in an area can be a core grievance.

6-122. Extremists that are very conservative can be as revolutionary in intent as other groups that are considered very liberal. Intents may include replacing forms of legitimate government with varied types of authoritarian rule. Socialism or variants of communism and/or other forms of totalitarianism can exist in many forms. However, ideology can range the political variable from fascist or other totalitarian intentions to an opposite political perspective of anarchism. As regimes fail to demonstrate its value to a subjected population, a population may gradually and secretly plan and act for change. Religious fundamentalism or extremist viewpoints can become a core grievance. Pervasive and desperate poverty often fosters and fuels widespread public dissatisfaction in a relevant population. Young people without jobs or hope are ripe for recruitment into an irregular force. Lack of essential services in a population complicates an OE and its effective governance. Examples of these essential needs are availability of food and potable water, credible law enforcement, emergency services, electricity, shelter, health care, schools, transportation, and sanitation programs for trash and sewage. Stabilizing a relevant population requires meeting these basic social and civil needs. Frustration and a feeling of disenfranchised existence can sway a relevant population to support an irregular force that promises change and improved living conditions.

ADAPTIVE ORGANIZATIONAL TACTICS AND TECHNIQUES

6-123. In an era of sophisticated weapon systems and rapid international deployment of military forces, incidents demonstrate the effectiveness of simple tactics conducted efficiently by small irregular forces. Devastating attacks can be conducted by a small unit or cell of irregular force terrorists. Operating in two- or three-person teams, they could simultaneously attack separate locations. The terrorists would be armed primarily with small arms, hand grenades, and improvised explosive devices. The attacks and eventual destruction of the terrorists could extend to hours or days of horrifying murder and mayhem. A methodical containment and room-by-room clearance of historic and multi-story buildings in a large metropolitan complex would be covered by global media outlets throughout the incident. An irregular force conducts operations with the expectation of such publicity to spotlight its irregular force agenda and embarrass the governing authority that it defies.

VERSATILE ORGANIZATIONAL AFFILIATIONS

6-124. Irregular forces will normally start as small entities with modest operational capabilities. As capabilities develop, plans may focus on acquiring improved capabilities across the functions of intelligence, logistics, communications, media affairs, and direct action. Building capacity in order to demonstrate commitment and capability is progressive. Temporary setbacks in plans and actions require flexibility to adjust timelines and adapt support networks. These networks can be as local as a neighborhood information-intelligence collection effort to a disciplined outreach for assistance from sources external to the immediate community or region.

6-125. Organizational versatility can include units or cells focused on acts of terrorism or other actions and organizations which are difficult to distinguish from civil crime. The increasing role of criminal activity in financing an agenda, either in partnership or competition with traditional criminal activities, can be a condition used by irregular forces. These enterprises include drug trafficking and human smuggling, fraud, tax evasion, counterfeiting, money laundering, and theft. Some activities are associated with terrorist's evolving capabilities for false documents production and concealment of money transactions for their operational purposes.

6-126. Irregular forces and criminal organizations can be closely related for mutual benefits. Mega-cities are expanding in countries with poor services and weak governance. Bases and operations in rural and urban environments will increase where law enforcement is often ineffective and/or corrupt. Rampant unemployment and dissatisfaction creates a productive recruiting ground and operating environment for irregular forces promoting grievances acknowledged by large segments of the population. Many of these city areas have materiel, communication, and transport capacities for irregular forces to use, and a potentially huge base of sympathizers and recruits.

6-127. A development related to this is the emergence of regions where governments exercise marginal control of geographic areas within their sovereign territory. Control is imposed by sub-state actors that can span criminal organizations, militias, guerrillas, insurgents, and terrorists. Government forces confront violent and capable irregular forces that often resort to open coercion and terrorism of officials, police officers, military forces, their families, and the general citizenry. In addition to overt intimidation of a populace, irregular forces exploit the vulnerabilities of new technologies to attack and have a great deal of flexibility in their use of new technology. They have the advantage of only needing to attack or neutralize specific systems or capabilities, and can concentrate fiscal expenditures on specialized counter-technology to protect their criminal systems. Nonetheless, irregular forces can often neutralize advanced systems or capabilities through the use of simple and unconventional techniques such as a suicide bomber.

EXPANDED TRANSNATIONAL AFFILIATES

6-128. Transnational actions and terrorism presents a global challenge. Some irregular forces seek to conduct major operations that cause large scale, maximum casualty impact. Globalization removes the perceived security that national borders and geographic distance from adversaries and enemies once indicated. Commerce and finance is international, travel is international, and society in general is a much more international community. Operating beyond the organizational reach of an opponent can provide the physical space or fiscal and media affairs support to enhance an irregular force purpose.

6-129. Irregular forces can operate in and among this international reach and develop transnational capabilities through a loose affiliation of clandestine networks, inserted sleeper or active cells within geographic regions, and adaptability to constantly shift and change in organizational form. When extremist ideology is the primary motivation, the commitment is often absolutist with no allowance for compromise and seeks no negotiation. Violence signals a committed path to conditions with no restrictions by social norms, laws, or values. Actions such as terrorism are often becoming more networks based. Irregular forces can encourage loosely organized, self-financed organizational structures. The motivation of terrorists appears to be based increasingly on theological extremes and ideological absolutes. International or transnational cooperation among some terrorists provides an improved ability to recruit members, develop fiscal support and resources, gain skills training and expertise, transfer technology, and when desired, political advice.

EMERGENT INDEPENDENT ADHERENTS

6-130. The presence of varied irregular forces and/or possible independent actors that separate from an originating organization can easily blur in organizations claiming responsibility for terrorism incidents. These actors can be at the individual level or small cells acting on a specified agenda that may be similar but distinct from more well known irregular force organizations. Independent actors may purposely minimize or preclude public announcement of their plans and action in order to improve their cell security.

6-131. Independent adherents are adjusting their financial operations to be self-sustaining in their activities. This independence from any external control is a value to an irregular force as it gains the impact of diverse and separate adherent actions at no real cost to his own organizational base. The facility with which groups can obtain and move funds, procure secure bases, and obtain and transport weaponry determines their own operational abilities and the level of threat that they pose to a common enemy. The international nature of finance, the integration of global economies, and the presence of terrorist adherents in the illegal economies of slaves, drugs, smuggling, human trafficking, counterfeiting, identity theft, and fraud have aided an independence from traditional sources of state-linked sponsorship and other interconnected support.

DEVASTATING FULL SPECTRUM WEAPONRY

6-132. Weapon lethality can be assessed in many ways but efficiency and effectiveness are two general means to determine the intent. Some weapons are quite simple and very deadly in effect. Lethality and overall effects are key considerations when assessed with types of action to obtain attention and reaction. The trauma of violent deaths as well as mass injuries and damage on a targeted population can be critical to the psychological effect and consequent actions or inaction by the affected population. Terrorism will continue to seek forms of indiscriminate violence. Acts of terror can also be targeted to specific people, groups, or capabilities. Terrorism is merging and combining with various other state and sub-state actors, further blurring the difference between criminals, rogue governments, and terrorists.

6-133. Ongoing conflicts display that terrorism attacks account for only a small fraction of civil violence but the high-profile nature of many terrorist operations can have a disproportionate negative impact. Conditions and changing dynamics in conflict can create a perception in a relevant population of unchecked violence and fear. Such perception can harden differing opinions, empower militias and vigilante groups, increase a middle-class exodus from a region, and disrupt confidence in a governing authority and its security forces. Terrorism plays a key role in much of this physical and psychological violence.

6-134. The means to cause mass casualties can be a highly sophisticated weapon system such as a weaponized military-grade nuclear device. However, mass destruction can also be caused by an improvised low-yield explosive weapon such as the 2001 attacks on the Twin Towers in New York City and the Pentagon. While purposeful mass casualty incidents may have appeared to be extraordinary events several decades ago, contemporary acts of terror surpass these former acts and demonstrate a profound impact on populations at the local, regional, national, and international levels. Emergent actions indicate that terrorism previously centralized and controlled by formal networks and organizations, is being conducted increasingly by loosely affiliated terrorists or groups that may generally identify themselves with an ideology or special purpose agenda. The threat can be foreign and domestic.

6-135. Proliferation of weapons of mass destruction is a particularly alarming issue. The projected effects amplify the danger and fear of a catastrophic attack. Some of these materials that could be used to construct some form of WMD are easily accessible in the public domain. Other potential capabilities are quite problematic to weaponizing materiel. The knowledge and technological means of specialists to produce WMD is a shadowy area of science, crime, and intrigue available to some terrorists.

6-136. The trend to exploit available simple and sophisticated technologies and the desire to cause mass casualties could be demonstrated in irregular force attacks that employ WMD such as—

- Chemical [sarin nerve agent] attacks in a metropolitan subway system.
- Biological [anthrax] attacks using contaminated letters in postal system.
- Low-yield explosive aerial bomb as in nearly simultaneous suicide attacks with large commercial jet aircraft into critical infrastructure.
- Low-yield explosive vehicular bomb as in a vehicle borne improvised explosive device attack on a Federal governing authority building and commercial business locale.
- Radioactive material placed in a public location for area contamination.
- Nuclear weapon detonation in a controlled environment as a demonstration of capability.

TARGETED ECONOMIC DISRUPTION

6-137. Modern, high-technology societies are susceptible to a concept of complex terrorism. Dependence on electronic networks in modern infrastructure, sometimes with questionable redundancy in systems, and concentrating critical assets in small geographic locales can present lucrative targets for the irregular force. Some emerging irregular forces wield effective power in failing states and view terrorism as an effective mode of conflict.

6-138. Critical infrastructure and support systems provide terrorists with a wide array of potential targets in land, maritime, cyber, and space environments that would directly affect economies on a regional and global scale. Common vulnerabilities may focus on oil refinery operations and overland and/or maritime shipment of oil products. Infrastructure for oil production has critical aspects in key areas of the world. Single points of failure in the infrastructure or denying critical services for a period of time could severely disrupt world economies.

GLOBALIZED MEDIA AFFAIRS

6-139. Exploiting mass media considers the value of effectively marketing an irregular force message through a nearly simultaneous global information environment. An overarching principle is to create and maintain international attention and impact. To gain international awareness and attention to grievances and issues, incidents must often be spectacular to attract mass media coverage. Effectiveness of INFOWAR will be measured by its ability to cause a dramatic impact of fear and uncertainty in a target population. Surprise and sustained violence may be normal against specified people representing elements of civil or military control and order of a governing authority with whom irregular forces are in conflict. Common citizens may also be targets in a culture of violence. Damage or destruction of community, regional, or national infrastructure and governance will be used to gain attention, provoke excessive reaction by host nation or coalition military forces, and attempt to alienate general population support of a governing authority.

6-140. Irregular force operations consider a desired media effect and plan for verbal or visual reporting coverage. Supporting events and interviews reinforce the desired message. These messages may present disinformation and false perspectives. Frequently, military reluctance to comment on ongoing operations in the media for operational security reasons can assist the irregular force. If no balanced information comes from official sources in a timely manner, the media may use the information readily available from the irregular force or a terrorist cell as a primary source for reporting an incident.

6-141. Future armed conflicts are more likely to be fought "among the people" rather than "around the people." Conflict in urban areas and populated locales is inevitable. Much of the world's population is already located in urban terrain. In the next two decades, estimates pose five billion of the world's eight billion people will live in cities. Many of the population hubs will be along coastlines throughout the world's regions. Irregular forces will attempt to blend into concentrated populations in order to marginalize detection of their recruiting, training, and staging initiatives; obtain support within a coerced or passive populace, and act at times and locations of their own selection. Media attention to a major incident will be immediate in coverage.

6-142. Rural areas and their inhabitants, depending on conditions in particular regions, will complement operations in urban centers. Irregular forces will evolve hybrid capacity of selective conventional military, paramilitary, and criminal organization capabilities. Rural irregular forces will operate in regions providing cover and concealment such as heavily forested or mountainous areas that hinder some forms of detection and interdiction. Notwithstanding, a rural orientation must lead back to urban centers as a presence and connection to the main population masses and the political infrastructure of governing authorities with whom irregular forces are in conflict.

THE SPECTER OF TERRORISM

6-143. The specter of irregular forces using terrorism on a society illustrates the value that such acts can provide to an irregular force that is often incapable of directly confronting an enemy with overwhelming combat power. The key factors of a successful campaign of irregular force terrorism will continue to be the long-term resolve and resilience of terrorists to continue the conflict until their aims are achieved.

6-144. The dedicated irregular force using terrorism must be adaptive and flexible to the periodic success of an enemy. When terrorism is disrupted with key irregular force leaders killed or captured, and functional capabilities are degraded or destroyed, other irregular force members, affiliates, and/or adherents are prepared to continue the struggle against a declared enemy.

6-145. A relevant population is a critical qualifying environment of nearly all irregular conflict. Irregular forces decentralize, network, and operate among a relevant population to overcome technological advantages of an enemy. The resilience of an irregular force and the expected debilitating effects of terrorism on an enemy require a determined long-term commitment from irregular force leaders to defeat an enemy. The massing of effects over time often requires a patient and decentralized approach to irregular conflict. The irregular force maintains a keen sense of cultural awareness and social understanding to establish a rationale for terror that is tolerated to a relevant population. The irregular force must obtain and sustain the passive and/or active support of the relevant population within which it operates.

6-146. An INFOWAR campaign must promote a resilient irregular force message of purpose and commitment to a relevant population. The objective of terrorism in support of an irregular force agenda must be recognized by a relevant population as a compelling outcome worth the amount of suffering and social disruption that terrorism and other acts by an irregular force may cause. Whether the ultimate aims of an irregular force are ideological, philosophical, and/or practical, terrorism will continue to be a vexing factor in future conflicts among and between state and non-state actors throughout the world.

Chapter 7

Functional Tactics

Insurgents and guerrillas, as part of the irregular OPFOR, may employ adaptive *functional* tactics. When planning a tactical action, an irregular OPFOR commander or leader determines what *functions* must be performed to accomplish the mission. Then he allocates functional responsibilities to his subordinates and synchronizes the effort.

Note. Since criminal elements do not normally have the ability to execute these functional tactics, the term *irregular OPFOR* in this chapter refers to insurgents and/or guerrillas.

FUNCTIONAL ORGANIZATION OF FORCES AND ELEMENTS

7-1. An irregular OPFOR commander or leader specifies the initial organization of forces or elements within his level of command, according to the specific functions he intends his various subordinates to perform. At brigade level (when that exists in guerrilla units), the subordinate units performing these functions are referred to as *forces*. At lower levels they are called *elements*.

Note. This functional organization provides a common language and a clear understanding of how the commander or leader intends his subordinates to fight functionally. Thus, subordinates that perform common tactical tasks such as disruption, fixing, assault, exploitation, security, deception, or main defense are logically designated as disruption, fixing, assault, exploitation, security, deception, or main defense forces or elements. Irregular OPFOR commanders and leaders prefer using the clearest and most descriptive term to avoid any confusion. When the irregular OPFOR operates in conjunction with an affiliated regular military force, there is the advantage that regular OPFOR commanders also use this common language.

7-2. The commander or leader organizes and designates various forces and elements according to their function in the planned offensive or defensive action. A number of different functions must be executed each time the irregular OPFOR attempts to accomplish a mission. The functions do not change, regardless of where the force or element might happen to be located. However, the function (and hence the functional designation) of a particular force or element may change during the course of a tactical action. The use of precise functional designations for every force or element involved in a particular tactical action allows for a clearer understanding by subordinates of the distinctive functions their commander or leader expects them to perform. It also allows each force or element to know exactly what all of the others are doing at any time. This knowledge facilitates the ability to make quick adjustments and to adapt very rapidly to shifting tactical situations. This practice also assists in a more comprehensive planning process by eliminating the likelihood of some confusion (especially on graphics) of who is responsible for what.

Note. A subordinate designated as a particular functional force or element may also be called upon to perform other, more specific functions. Therefore, the function of that force or element, or part(s) of it, may be more accurately described by a more specific functional designation. For example, a disruption force generally "disrupts," but also may need to "fix" a part of the enemy forces. In that case, the entire disruption force could become the fixing force, or parts of that force could become fixing elements.

7-3. The various functions required to accomplish any given mission can be quite diverse. However, they can be broken down into two very broad categories: action and *enabling*.

ACTION FORCES AND ELEMENTS

7-4. One part of the organization conducting a particular offensive or defensive action is normally responsible for performing the primary function or task that accomplishes the overall mission goal or objective of that *action*. In most general terms, therefore, that part can be called the *action force* or *action element*. In most cases, however, the higher unit commander will give the action force or element a more specific designation that identifies the specific function or task it is intended to perform, which equates to achieving the objective of the higher command's mission.

7-5. For example, if the objective of the action of a guerrilla company or a grouping of insurgent direct action cells is to conduct a raid, the element designated to complete that action may be called the *raiding element*. In a brigade-level offensive, a force that completes the primary offensive mission by exploiting a window of opportunity created by another force is called the *exploitation force*. In defensive actions, the grouping that performs the main defensive mission is called the *main defense force* or *main defense element*.

ENABLING FORCES AND ELEMENTS

7-6. In relation to the action force or element, all other parts of the organization conducting an offensive or defensive action provide *enabling* functions of various kinds. In most general terms, therefore, each of these parts can be called an *enabling force* or *enabling element*. However, each subordinate force or element with an enabling function can be more clearly identified by the specific function or task it performs. For example, an element that *clears* obstacles to permit an action element to accomplish a battalion's tactical task is a *clearing element*.

7-7. In most cases, irregular OPFOR tactical actions would involve one or more types of enabling forces or elements designated by their specific function. The most common types include—

- *Disruption force or element.* Disrupts enemy preparations or actions; destroys or deceives enemy reconnaissance; begins reducing the effectiveness of key enemy elements.
- *Fixing force or element.* Fixes the enemy by preventing a part of his force from moving from a specific location for a specific period of time, so it cannot interfere with the primary OPFOR action.
- *Security force or element.* Provides security for other parts of a larger organization, protecting them from observation, destruction, or becoming fixed.
- *Deception force or element.* Conducts a deceptive action (such as a demonstration or feint) that leads the enemy to act in ways prejudicial to enemy interests or favoring the success of an OPFOR action force or element.
- *Support force or element.* Provides support by fire; other combat or combat service support; or command and control (C2) functions for other parts of a larger organization.
- *Reconnaissance force or element.* Conducts specific reconnaissance task(s) tailored to the mission.

Note. In a defensive situation, there may be a particular unit or grouping that the irregular OPFOR commander or leader wants to be protected from enemy observation and fire, to ensure that it will still be available for further actions. This may be designated as the *protected force or element*.

FLEXIBILITY

7-8. The function of a particular force or element may change in the course of a tactical action. For example, a grouping originally designated as a reconnaissance element may locate an enemy unit that it has the combat power to engage. In that case, the element in question could become an assault element, ambush element, or raiding element.

Functional Tactics

TYPES OF OFFENSIVE ACTION

7-9. Insurgents and guerrillas can employ some of the types of offensive action also used by smaller tactical units of the regular OPFOR. Such actions can include—
- Ambush.
- Assault.
- Raid.
- Reconnaissance attack (guerrillas only).

Insurgent cells typically do not have sufficient combat power to conduct a reconnaissance attack. (See TC 7-100.2 for basic discussion of these offensive actions, as they are also conducted by the regular OPFOR.)

7-10. Irregular OPFOR leaders and commanders select the offensive action best suited to accomplishing their mission. Insurgent cells and small guerrilla units typically execute one combat mission at a time. Therefore, it would be rare for such a cell or unit to employ more than one type of offensive action simultaneously. However, irregular OPFOR organizations are dynamic and adapt very quickly to the situation. An offensive action may have to make use of whatever cell(s) or unit(s) can take advantage of a window of opportunity.

AMBUSH

7-11. An ambush is a surprise attack from a concealed position against a moving or temporarily halted target. In an ambush, the actions of the enemy determine the time, and the irregular OPFOR leader decides on the location. Similar to purposes used by regular military OPFOR, the irregular OPFOR can conduct ambushes to—
- Destroy or capture enemy elements, personnel, and/or designated very important persons.
- Secure supplies.
- Demoralize enemy military forces and officials of a governing authority.
- Delay introduction of international and/or enemy coalition assistance to a governing authority.
- Block enemy movements and/or logistics support.
- Canalize or restrict enemy movement.

7-12. The irregular OPFOR can use an ambush as a primary psychological tool in its information warfare (INFOWAR) activities. The psychological effects of ambushes can be enhanced by—
- Conducting recurring ambushes at known areas and/or points where enemy forces must travel.
- Changing the tempo or the number of ambushes to appear unpredictable.
- Attacking targets that were previously considered safe or had not been attacked.
- Using weapons with range capabilities previously not used in an area of conflict.
- Increasing weapons and/or demolitions effects against particular targets.

7-13. A common tactic is to conduct an ambush as a means to set up ambush(es) of enemy forces that respond to the original ambush. Multiple and nearly simultaneous ambushes can be conducted along likely avenues of approach to the area of the initial ambush. Ambushes may also target enemy medical treatment and evacuation assets, when irregular OPFOR commanders or leaders decide to not comply with international conventions and law of war norms that regular military forces use. The destruction of means to evacuate and treat wounded can instill a sense of tentativeness in enemy soldiers because they realize that, should they become wounded or injured, medical help may not be forthcoming.

7-14. Attacking known points of enemy weakness is a fundamental planning consideration for the irregular OPFOR. Correspondingly, the irregular OPFOR avoids enemy strength.

7-15. Surprise and overwhelming massed firepower at a specific place and time provides an expectation of tactical success for the irregular OPFOR. Factors that complement tactical surprise and massed firepower are—
- Detailed plans and rehearsals.
- Selection of ambush positions.

- Rapid and violent conduct of the ambush.
- Disciplined withdrawal of irregular OPFOR elements from the ambush site.

Functional Organization for an Ambush

7-16. An ambush force is typically organized into three types of elements: the *ambush element*, *security element*, and *support element*. There may be more than one of each element (see figure 7-1).

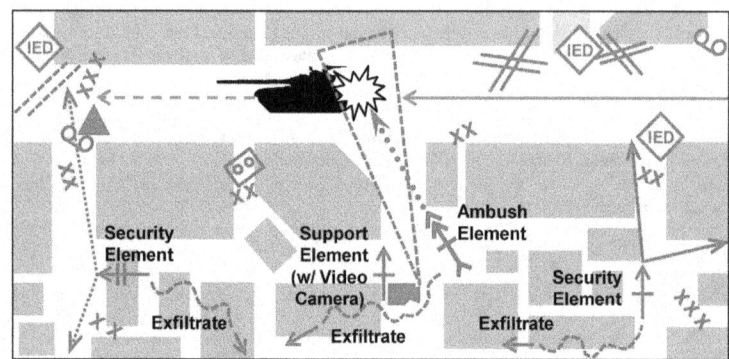

Figure 7-1. Insurgent ambush (example)

Ambush Element(s)

7-17. The *ambush element* has the mission of attacking and destroying enemy elements in kill zone(s). Other tasks may include capturing personnel and/or recovering supplies and equipment.

Security Element(s)

7-18. The *security element* has a mission to provide early warning to irregular OPFOR elements of any enemy presence that might disrupt the ambush. Another task can be to protect the ambush element from becoming decisively engaged by enemy forces before, during, or after the ambush.

Support Element(s)

7-19. The *support element* can include direct and/or indirect fires and provides general support to improve success of the ambush. The insurgent leader or guerrilla commander typically commands and controls the ambush from the support element. However, he will position himself where he can best command and control.

Executing an Ambush

7-20. There are three types of ambushes based on the desired mission effects—annihilation, harassment, or containment. The irregular OPFOR conduct ambushes with a particular purpose that often supports a larger tactical action.

Annihilation Ambush

7-21. The purpose of an annihilation ambush is to destroy an enemy force within a designated kill zone. In addition to massed direct fires, the irregular OPFOR often increases the lethality of a kill zone with indirect fires, manmade obstacles, mines, and/or improvised explosive devices (IEDs) to halt, contain, and kill the enemy force in the kill zone (see figure 7-2).

Note. For guerrilla forces, annihilation ambushes in complex terrain, including urban environments, often involve task-organized hunter-killer (HK) teams.

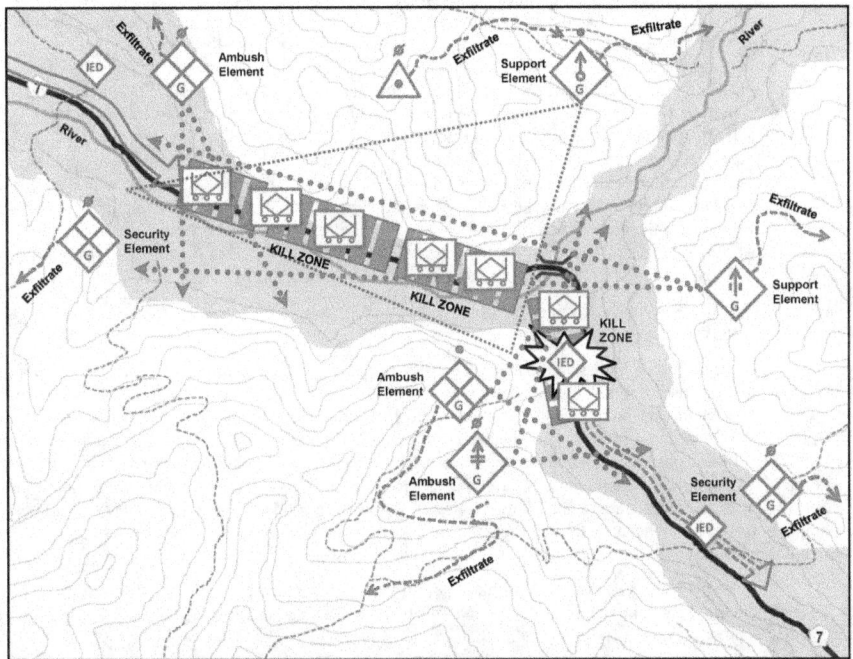

Figure 7-2. Guerrilla ambush (example)

7-22. Irregular OPFOR commanders and leaders may be willing to accept decisive engagement with the enemy in this type of ambush. An annihilation ambush typically emphasizes tactical tasks to—

- Block.
- Contain.
- Destroy.

7-23. The ambush and support elements normally remain in their fighting positions until the enemy in the kill zone is rendered combat ineffective. The intent is to destroy enemy personnel and equipment within the kill zone with concentrated firepower.

7-24. Once the enemy is destroyed, the ambush element can secure the kill zone and eliminate any remaining enemy in the kill zone. The support element provides overwatch protection to the ambush element when the ambush element is directed to search the destroyed enemy force and equipment for information and/or intelligence. Weapons and materiel can be seized by the ambush element for future irregular OPFOR tactical actions.

7-25. The security element remains in fighting positions to ensure early warning, isolate a kill zone, and prevent any enemy from escaping the kill zone. Once the ambush element clears the kill zone, the ambush force withdraws from the ambush area. The ambush element withdraws first and is followed by the support element. The security element is the last element to depart the kill zone area and delays or blocks any rapid response of enemy forces that attempt to pursue irregular OPFOR elements as they depart the kill zone area. Depending on the size of the ambush force, the elements typically reassemble at a predetermined location and time at a safe house or safe haven.

> *Note.* An irregular OPFOR ambush could employ security elements to provide early warning and/or isolate a series of kill zones on a known convoy route of the enemy. Restrictive natural terrain and manmade features are reinforced with IEDs to disrupt and contain an enemy force in the kill zones. A simple ambush technique is to employ a decoy IED that is observable on an enemy force route. Once enemy forces halt to investigate the potential of an IED detonation, the ambush force initiates the actual ambush with the simultaneous detonation of IEDs directed into the kill zone where enemy soldiers and vehicles are expected to halt. Lead and trail vehicles are initially the primary targets for massed direct fires and destruction. When they are destroyed, the ambush and support elements shift direct and indirect fires from both ends of the enemy column toward the center of the contained enemy forces. The enemy convoy is destroyed with massed overlapping direct and indirect fires. Ambush elements and designated support elements exfiltrate from the area while security elements provide rear security and an all-arms air defense capability against any enemy response forces. On order, security elements also exfiltrate from the ambush site and rendezvous with other guerilla elements at a safe haven.

Harassment Ambush

7-26. The purpose of a harassment ambush is to disrupt routine enemy activities, impede the enemy's freedom of movement, and/or create a negative psychological impact on enemy personnel. The irregular OPFOR may choose to conduct a harassment ambush when the enemy has superior combat power and destruction of an enemy force is not feasible as in an annihilation ambush. This type of ambush does not require the use of obstacles to keep the enemy in the kill zone but can include terrain reinforced into a kill zone with manmade obstacles, mines, and/or IEDs to halt and/or contain the enemy force for a limited period of time. Compared to an annihilation ambush, the irregular OPFOR typically conducts a harassment ambush at a greater distance from the enemy in a kill zone. A harassment ambush often considers the maximum effective range of its weapons when massing firepower.

7-27. The irregular OPFOR does not normally accept decisive engagement with the enemy in this type of ambush. A harassment ambush typically emphasizes tactical tasks that can include—

- Disrupt.
- Delay.
- Defeat.

7-28. The ambush and support elements are often combined to provide more effective control of fires throughout the kill zone. This combination is especially useful when the kill zone is quite wide and/or extends for a long distance. The security element provides early warning of any enemy forces conducting reconnaissance prior to the ambush and/or enemy forces attempting to respond to the ambush.

7-29. Once the irregular OPFOR commander or leader determines that the ambush has achieved the desired effects, he directs the ambush and support elements to withdraw along designated routes. The security element continues to report on enemy activity in the kill zone area and any attempt of enemy forces to pursue. The irregular OPFOR will not become decisively engaged by enemy forces and often emplace mines and/or IEDs to delay enemy pursuit.

7-30. Repeated harassment ambushes against the enemy can—

- Cause the enemy to allocate a disproportionate amount of forces to security tasks which affect other enemy force missions and potentially create enemy vulnerabilities.
- Create a negative psychological effect upon enemy soldiers and leaders, and officials of a governing authority with which the irregular OPFOR is in conflict.

Functional Tactics

Containment Ambush

7-31. A containment ambush is a security task that is usually part of a larger tactical action. This type of ambush can prevent the enemy from using an avenue of approach or interdicting another tactical action such as a raid or another ambush.

7-32. The ambush element can be directed to secure a kill zone, but this task is not necessarily required for mission success. The support and security elements perform the same functions as those described in an annihilation ambush. Obstacles are an integral part of a successful containment ambush. The commander or leader determines if his relative combat power compared to enemy forces is adequate to conduct a containment ambush. The fact that containment may require the irregular OPFOR elements to remain in an ambush site for an extended period places those elements in danger of being fixed and defeated by enemy reinforcements.

7-33. The ambush force will normally not accept decisive engagement with the enemy in this type of ambush. However, it can be directed to accept decisive engagement in support of a larger irregular OPFOR action. A containment ambush typically emphasizes related tactical tasks that can include—

- Contain.
- Fix.
- Delay.
- Defeat.

Command and Control of an Ambush

7-34. The commander or leader of the ambush force normally positions himself with the support element and designates a subordinate leader to move and maneuver with the ambush element. However, the ambush force commander or leader locates himself where he can best command and control the ambush.

7-35. Urban and rural complex terrain provides several tactical advantages to irregular OPFOR ambush, security, and support elements. Operating among indigenous citizens in an urban area or other complex terrain can be used to—

- Observe enemy forces along known canalized routes or areas of reconnaissance and/or avenues of approach or directions of attack.
- Provide for easily camouflaged irregular OPFOR reconnaissance and surveillance activities.
- Provide covered and/or concealed irregular OPFOR routes into and out of the ambush kill zone area.
- Improve irregular OPFOR ambush, security, and support positions with cover, concealment, and camouflage of the natural and manmade tactical environment.
- Encourage deception activities in a relevant civilian population against enemy forces and a governing authority.
- Encourage techniques that employ overlapping direct fires from multiple directions into a designated kill zone.

Support of an Ambush

7-36. An ambush typically requires several types of support. These can include reconnaissance, fire support, air defense, engineer-like capabilities, logistics, and INFOWAR. Covert or overt assistance may also be provided from external sources such as special-purpose forces (SPF) of another state.

Reconnaissance

7-37. Reconnaissance is critical to a successful ambush and is continuous in the objective area in order to confirm and/or adjust information collection and intelligence previously collected and analyzed. The irregular OPFOR uses active supporters in the relevant population to observe and report on enemy activities at the planned objective area in order to select the best terrain on which to locate irregular OPFOR positions.

7-38. The irregular OPFOR often has sufficient time in a local community and in vicinity of the objective to observe and interact with the relevant population. Posing as innocent civilians or coercing local civilians and/or civic leaders, irregular OPFOR reconnaissance and surveillance reports combine with the reports from active supporters. Assessing and analyzing these reports assist the commander or leader in finalizing his ambush plan. Once infiltration and exfiltration routes are planned, the irregular OPFOR maintain these routes under constant surveillance prior to and during the ambush. Secrecy of irregular OPFOR locations and activities is essential to tactical survival.

7-39. The irregular OPFOR will closely monitor—
- Routines of enemy forces selected as the target.
- Enemy use of weapons and equipment, crew duties, and teamwork.
- Lapses in local security measures among groupings of enemy soldiers and vehicles.
- Routes and response time of enemy quick reaction forces during prior ambushes or irregular OPFOR ruses.
- Cooperation of governing authority law enforcement and paramilitary units with enemy forces.
- Medical treatment and evacuation norms of enemy forces.

Fire Support

7-40. Fire support is typically in the support element. In most ambushes, support elements in proximity to ambush elements provide supporting direct fires from light, medium, or heavy machineguns and/or antitank grenade launchers (ATGLs). However, some support elements may also provide indirect fires from mortars and rockets. Fires can also support security elements, if necessary. The irregular OPFOR emplaces fire support systems with the intention of quickly withdrawing them at the conclusion of the ambush.

7-41. Artillery from an affiliated regular military force can augment fires organic to the irregular OPFOR ambush force. Such artillery support can provide additional fires into the kill zone, illumination over it, or smoke to permit withdrawal.

Air Defense

7-42. Capabilities for air defense during an ambush may be limited to an all-arms air defense concept, using the small arms and direct fire weapons with the ambush force. However, guerrilla battalions typically have a limited man-portable air defense system (MANPADS) capability in their weapons company. If allocated to an ambush force, these MANPADS would likely be located in security element(s).

> *Note.* Air defense in the irregular OPFOR can also be improved by state and non-state sponsors providing sophisticated air defense weapons, technologies, training, and logistics support. Other opportunities may arise when insurgents or guerrillas in armed conflict capture or acquire sophisticated air defense weapons. In either case, clandestine state or non-state agents and/or technicians can provide technical support to ensure the effective use of the weapon systems. Examples of state-of-the-art air defense systems include shoulder-fired MANPADS and/or other air defense missiles with detection and tracking systems mounted on wheeled or tracked vehicles.

Engineer-Like Capabilities

7-43. Mobility and countermobility support often depends on insurgents or guerrillas with specialized skills and expertise from their civilian occupations or previous military experience. Guerrilla units include sappers, who are not engineers but can perform some engineer-like functions. Covert or overt assistance may also be provided from external sources such as SPF of another state.

Logistics

7-44. Logistics are prepared as caches or supported from safe houses and havens as part of detailed planning and rehearsals. The ambush force typically moves from a secured location with everything it needs to complete the mission. In those rare situations that require a multi-day hide prior to executing the ambush, the ambush force will have to move with its own extra life support. Resupply of the ambush force would significantly increase the chances of its detection and defeat its purpose.

INFOWAR

7-45. INFOWAR activities can support ambushes by concealing the intended action through deception and information protection. An INFOWAR campaign may use successful ambushes to demonstrate the progressive failure of an enemy force and/or governing authority. INFOWAR support of an ambush can temporarily and psychologically isolate the enemy force. (See appendix A for additional information on irregular OPFOR INFOWAR capabilities.)

ASSAULT

7-46. An assault is an attack that destroys an enemy force through firepower and the physical occupation and/or destruction of his position. An assault is a basic form of irregular OPFOR tactical offensive combat. Therefore, other types of offensive action may include an element that conducts an assault to complete the mission. However, that element will typically be given a designation that corresponds to the specific mission accomplished. For example, an element that conducts an assault in the completion of an ambush would be called the ambush element.

Functional Organization for an Assault

7-47. The insurgent cell(s) or guerrilla unit(s) conducting an assault constitute an assault force. The assault force typically is organized into three types of elements:
- Assault element.
- Security element.
- Support element.

There may be more than one of each of these types of element.

Assault Element

7-48. The *assault element* is the action element. It maneuvers to and seizes the enemy position, destroying any forces there.

Security Element

7-49. The *security element* provides early warning of approaching enemy forces and prevents them from reinforcing the assaulted enemy unit. Security elements often make use of terrain choke points, obstacles, ambushes, and other techniques to resist larger forces for the duration of the assault. The commander or leader may (or may be forced to) accept risk and employ a security element that can only provide early warning that is not strong enough to block or delay enemy reinforcements. This decision is based on the specific situation.

Support Element

7-50. The *support element* provides the assault element with one or more of the following:
- C2.
- Combat service support (CSS).
- Supporting direct fire (such as small arms, grenade launchers, or ATGLs).
- Supporting indirect fire (such as mortars or rockets).
- Mobility support.

7-51. The assault force commander or leader typically commands and controls the assault from the support element. However, he will position himself where he can best exercise C2.

Executing an Assault

7-52. An assault is a rapid and violent action that can have significant and even decisive effects. However, a simple direct assault has a very low chance of success without some significant mitigating factors. Decisive assaults are characterized by—

- Isolation of the objective (enemy position) so that it cannot be reinforced during the assault.
- Early warning of any approaching enemy reinforcements and/or other security measures by the security element.
- Effective suppression of the enemy by the support element prior to the assault element maneuvering on the enemy position.
- Violent fire and maneuver into and through the enemy position.

7-53. The assault element maneuvers from its assault position to the objective and destroys the enemy located at the objective. It can conduct attack by fire, but this is often not an optimal method and should be used only when necessary. Typical tactical tasks of the assault element are—

- Clear.
- Destroy.
- Seize.

7-54. The irregular OPFOR normally does not assault to secure, since this task indicates an intention to prevent the loss of an objective to subsequent enemy reaction. Any occupation of an objective is typically temporary to minimize the ability of an enemy force to mass overwhelming combat power against the irregular OPFOR. Speed of execution and surprise are critical to an assault. (See figures 7-3 and 7-4 for examples of assaults.).

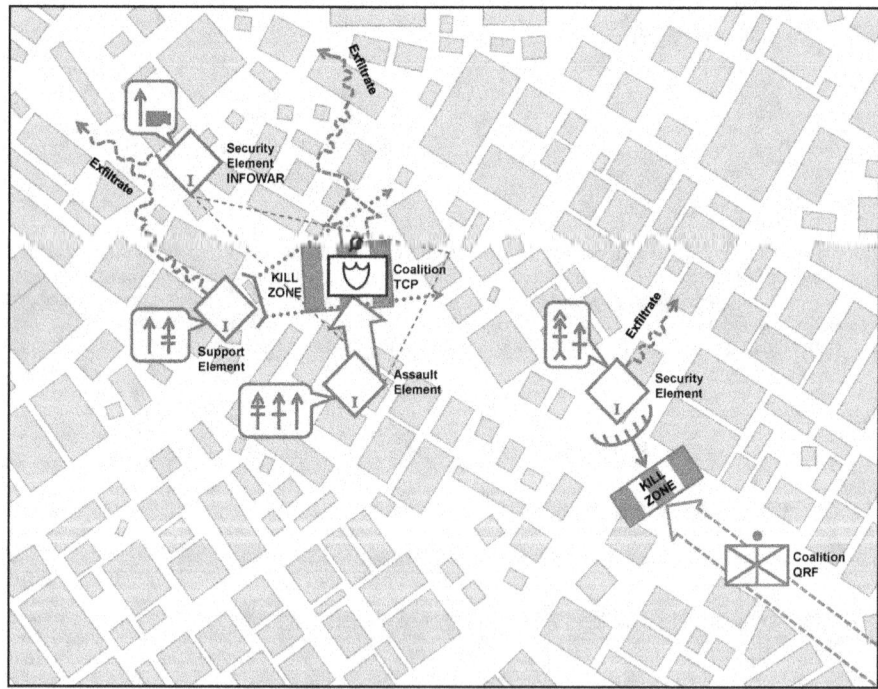

Figure 7-3. Insurgent assault (example)

Functional Tactics

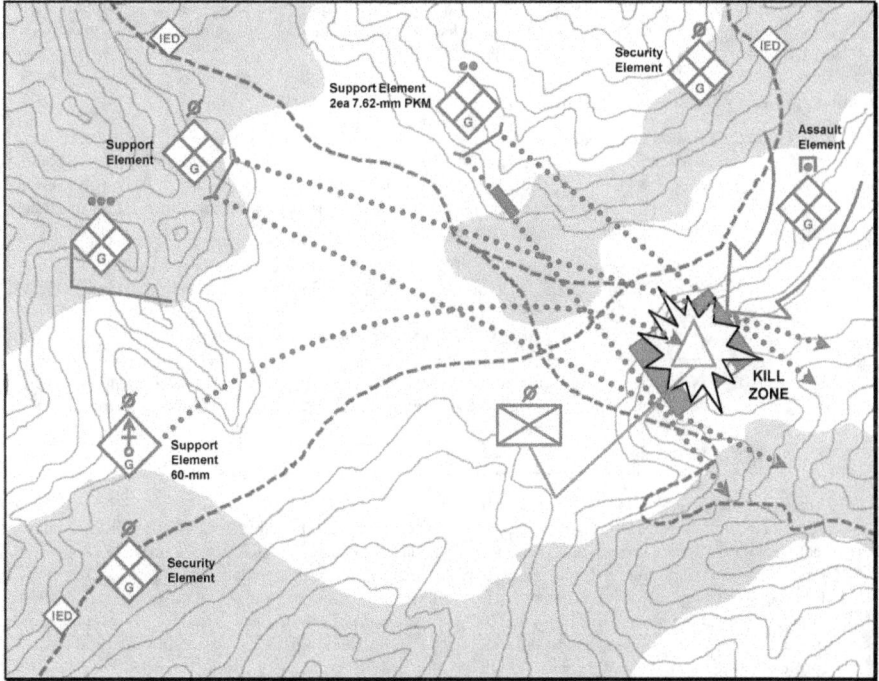

Figure 7-4. Guerrilla assault (example)

7-55. The security element is equipped and organized to detect enemy forces that may be able to react to an assault on an objective. While the assault is in progress, security tasks may include the requirement to isolate the objective from any reinforcement by enemy forces, with tasks such as block, contain, or delay. The security element may be directed to conduct similar tasks that allow the assault and support elements to exfiltrate.

> *Note.* A simple, effective, and successful assault technique employed often by the irregular OPFOR is to surprise the enemy by focusing enemy attention in one direction and then assaulting from a different direction in a nearly simultaneous action with massed firepower and maneuver on the objective. Support element(s) shift and/or lift initial small arms fire across the objective as the assault element assaults through the objective. Security element(s) prepare to ambush any enemy response forces and provide early warning to other elements. The assault element is already exfiltrating from the objective when the assault force commander or leader gives the order to support and security element(s) to exfiltrate from the area. A security element with an INFOWAR team can record the successful assault and relay the video and audio coverage to an intermediary for delivery to local media outlets. The INFOWAR team can publish its videotape with added narrative from a higher headquarters spokesperson on an Internet website often within hours of the assault.

7-56. The commander or leader of the assault force typically exercises C2 from within the support element. He can also lead the assault element when he determines that his personal presence in the assault is critical to mission success.

7-57. The support element controls all combat support (CS) and CSS functions as well as any supporting fires. Tasks expected of support elements in the assault are normally attack by fire and/or support by fire. These direct and/or indirect fires are often intended to divert the enemy's attention during the maneuver by the assault element.

Command and Control of an Assault

7-58. Command and control of the assault masses irregular OPFOR capabilities in time and space for rapid and violent attack on a selected objective. The irregular OPFOR normally plans an assault in detail to improve tactical execution with surprise and deception and in order to achieve temporary superior combat power against an enemy force. However, the irregular OPFOR will often take advantage of an unexpected opportunity as it occurs in order to assault an enemy force.

7-59. Fighting in complex terrain can be a significant tactical advantage for the irregular OPFOR. In urban areas the irregular OPFOR can use the civilian community and its infrastructure for shielding and obstacles against enemy forces. The irregular OPFOR may decide to not comply with international conventions and law of war restrictions that apply to regular military forces and governing authorities.

> *Note. Complex terrain* is a topographical area consisting of an urban center larger than a village and/or of two or more types of restrictive terrain or environmental conditions occupying the same space. (Restrictive terrain or environmental conditions include but are not limited to slope, high altitude, forestation, severe weather, and urbanization.)

7-60. Channelized corridors of urban traffic networks and the vantage points of multi-story buildings and/or surface or subsurface infrastructure can be demoralizing to an enemy force attempting to counter the irregular OPFOR. People and facilities in urban areas provide cover and concealment to the irregular OPFOR. Beyond that, enemy forces often operate within a relevant population that they do not want to alienate due to excessive civilian casualties, restrictions, and/or damage to their facilities and livelihoods. In comparison, rural complex terrain may have fewer people within designated areas but can provide similar cover and concealment advantages to the irregular OPFOR. The civilians that reside in such rural areas can be influenced to support the irregular OPFOR involuntarily or voluntarily through an effective INFOWAR campaign.

Support of an Assault

7-61. Support of an assault normally includes reconnaissance, logistics, and INFOWAR. When required for a particular mission, capabilities such as fire support and air defense can be added to the elements conducting an assault.

Reconnaissance

7-62. Reconnaissance effort for an assault is continuous in the objective area in order to confirm and/or adjust information collection and intelligence previously collected and analyzed. The irregular OPFOR uses active supporters in the relevant population to observe and report on enemy activities at the planned objective area. Insurgents or guerrillas are often positioned in the local community and in vicinity of the objective posing as innocent civilians conducting normal commercial or social actions. These reconnaissance and surveillance reports, combined with the reports from active supporters, assist the commander or leader of the assault force in finalizing his assault plan to shape, assault, and exfiltrate. Once infiltration and exfiltration routes are planned, the irregular OPFOR maintains these routes under constant surveillance prior to and during the assault. Secrecy of irregular OPFOR locations and activities is essential to tactical survival. Locations of keen interest for reconnaissance and surveillance include—

- Caches.
- Infiltration routes.
- Assault position.
- Support position.
- Objective.

- Exfiltration routes.
- Safe houses.

Fire Support

7-63. The primary mission of fire support in an assault is to suppress the objective and protect the advance of the assault element. Fire support assets are typically part of the support element(s). In most ambushes, support elements in proximity to ambush elements provide supporting direct fires from machineguns and/or ATGLs. However, the support element(s) can also include indirect fire weapons such as mortars and rockets allocated to the assault force.

Air Defense

7-64. The typical purpose of air defense support to an assault is to prevent enemy air power from influencing the action of the assault element. All three elements of an assault typically employ the concept of all-arms air defense. If specialized air defense weapons (such as MANPADS) are available, they could be used in any of the elements, but are least likely to be found in the assault element. The security element provides early warning of enemy aerial response to the assault and may try to destroy the enemy aircraft. The support element provides overwatch of the assault element and the objective. (See note under Air Defense in Support of an Ambush regarding additional air defense assets that may be available.)

Logistics

7-65. Logistics support for an ambush is similar to that for an ambush (see above). The support element is responsible for CSS.

INFOWAR

7-66. INFOWAR support of an assault considers the rapid and violent nature of an assault and the intention to temporarily and psychologically isolate the enemy force. Isolation of the enemy may also use physical means such as simultaneous assaults on multiple objectives to overload the enemy's ability to respond and/or effectively reinforce an enemy force at a particular irregular OPFOR objective. (See appendix A for additional information on irregular OPFOR INFOWAR capabilities.)

RAID

7-67. A raid is an attack against a stationary target for the purposes of its capture or destruction that culminates in the withdrawal of the raiding force to safe territory. Raids are usually small-scale attacks that use surprise and combat power to successfully accomplish the purpose of the mission. Sudden violence characterizes most raids and may be conducted to secure information, materiel, or individuals and can also be used to confuse or deceive an enemy. A raid concludes with a planned withdrawal upon completion of the assigned mission.

7-68. Raids can be used to—

- Destroy or damaging key systems or facilities (such as command posts, communication facilities, supply depots, radar sites).
- Seize hostages and/or prisoners.
- Rescue insurgents, guerrillas, and /or active supporters being detained and/or imprisoned.
- Destroy, damage, or capture supplies or lines of communication.
- Obtain or denying critical information to the enemy.
- Support INFOWAR actions that distract attention from other irregular OPFOR actions, keep the enemy off balance, and/or to cause the enemy to deploy additional units to protect critical sites.

Functional Organization for a Raid

7-69. The size and configuration of a raiding force depends upon its mission, the nature and location of the target, and the enemy situation in the objective area. Examples various raiding forces and their missions could include—

- A small insurgent cell or guerrilla HK team attacking an isolated voting station or a portion of unprotected railroad track.
- A larger raiding force attacking an enemy checkpoint, convoy route, or a large supply depot.
- A larger insurgent cell, group of cells, or task-organized guerrilla unit attacking an enemy combat outpost that is attempting to interdict irregular OPFOR movements and control in a geographic area.

7-70. Regardless of size and specific capabilities, a raiding force typically consists of three elements: raiding, security, and support. However, a raiding force may employ other functional elements such as a fixing element or breaching element. It may also obtain advice and direct assistance from SPF teams and or regular military forces that are in conflict with the same enemy force or governing authority.

Raiding Element(s)

7-71. The raiding element executes the primary task of the raid. That is to destroy or seize the objective of the raid. In some situations, the raiding element moves physically into the objective, and in other cases it is able to accomplish the raiding task from a distance. Other elements of a raid support and/or protect the raiding element while it approaches, enters, and departs the objective.

Security Element(s)

7-72. The security element in a raid is primarily focused on enemy containing enemy security forces, blocking enemy response forces, and/or fixing enemy escape from the objective area. Any of these tasks are usually conducted for a limited time period in support of the raid objective. The security element is often equipped and organized to detect enemy forces in the vicinity of the objective and prevent them from alerting enemy forces at the objective. Insurgents or guerrillas may infiltrate into the objective area and position themselves posing as civilians until the time of mission execution.

7-73. The task of a security element is to occupy enemy security and response forces and fix these enemy forces so that they cannot react to the raiding element. Security elements deploy to locations where they can deny the enemy freedom of movement along any ground or air avenues of approach that can reinforce the objective or interfere with the raid mission. Insurgents and guerrillas employ an all-arms air defense concept that uses all available weapons to disrupt and/or defeat enemy aircraft. Any additional air defense assets are most likely found in the security element(s).

7-74. Covering the withdrawal of the raiding element with a designated level of rear security, the security element typically does not allow itself to become decisively engaged. The size of the security element depends upon the size of the enemy's estimated capability to intervene and disrupt the raid.

Support Element(s)

7-75. The support element in a raid serves several enabling functions that assist in setting the conditions for success of the raid. This support may take several forms. The support element provides fire support, logistics, mobility and countermobility actions, and INFOWAR support to the raiding and security elements.

7-76. The commander or leader of the raiding force normally commands and controls the raid from within the support element. However, he will position himself where he can best command and control the raid.

7-77. Critical support element tasks are often executed immediately prior to conduct of the raid and/or facilitate its execution. Tasks that assist the raiding element(s) to achieve their objective can include—

- Breaching and removing obstacles to the objective.
- Conducting diversionary actions.
- Providing fire support.

Executing a Raid

7-78. Irregular OPFOR leaders and commanders plan for a rapid and violent execution of a raid. They do not intend to be decisively engaged with the enemy. The security and support elements normally remain in their fighting positions unless a task requires an element to accompany the raiding element into the objective. A breaching element may be required to clear a lane or lanes in enemy defenses and pass the raiding element(s) through to the objective.

7-79. When the intent is to destroy enemy personnel and equipment within the kill zone, concentrated firepower may be able to accomplish the task without physically entering the objective. For example, a guerrilla unit could use indirect fire weapons to raid an enemy site as an attack by fire with no intention of entering the objective site (see the examples in figure 7-5 and 7-6 at pages 7-16 to 7-17).

7-80. If the raiding task includes seizing individuals and/or equipment, the raiding element temporarily secures the objective and seizes designated equipment, individuals, and/or other materiel. The support element provides protection to the raiding element when the raiding element searches the objective for information and/or intelligence. The raiding element can seize weapons and other materiel for future irregular OPFOR tactical actions.

Command and Control of a Raid

7-81. A raid is conducted by elements that are often autonomous from other irregular OPFOR or regular OPFOR cells or units but can be coordinated in actions to support a common purpose. A raid is not necessarily associated with actions being conducted concurrently by larger OPFOR organizations in the same area. Irregular OPFOR raids are typically conducted by small insurgent cells or guerrilla units at the tactical level of conflict.

7-82. The commander or leader of the raiding force normally positions himself with the support element and designates a subordinate leader to move and maneuver with the raiding element. However, the raiding force commander or leader locates himself where he can best command and control the raid and may maneuver with the raiding element into the objective when appropriate.

Support of a Raid

7-83. A raid typically requires several types of support. These types of support can typically include reconnaissance, fire support, air defense, engineer-like capabilities, logistics, and INFOWAR.

Reconnaissance

7-84. The primary task of reconnaissance and surveillance in a raid is to collect information and intelligence on the target of the raid and monitor all activities in and near the objective. Reconnaissance also identifies the locations of response forces and their expected response routes to the objective during a raid.

7-85. Reconnaissance elements confirm and/or adjust information collection collected previously in order to determine current intelligence. Insurgents and guerillas are often members of a local community and easily interact with the relevant population. The irregular OPFOR uses active supporters in the relevant population to observe and report on enemy activities at the objective area in order to select the best terrain in which to locate irregular OPFOR positions. Infiltration and exfiltration routes are planned and rehearsed, and security elements maintain these routes under constant surveillance prior to and during the assault.

7-86. In preparation for a raid, the irregular OPFOR will closely monitor—
- Patterns of enemy force activities selected as the target.
- Lack of discipline in enemy soldier use of weapons and equipment, crew duties, and teamwork.
- Lapses in enemy local security measures in the objective area.
- Availability of enemy quick reaction forces.

Raid Examples

Insurgent Raid

Insurgents could use a raid to rescue an insurgent leader who is being held in a governing authority detention facility. A suicide vehicle-borne IED detonates at the main gate to breach barriers and kill, wound, or daze internal security force (ISF) guards in the immediate vicinity. Concurrently, support elements use small arms and ATGL fire to kill and/or contain other guard forces at a separate gate and small barracks. Security elements detect and ambush enemy reaction forces coming to the aid of the detention facility.

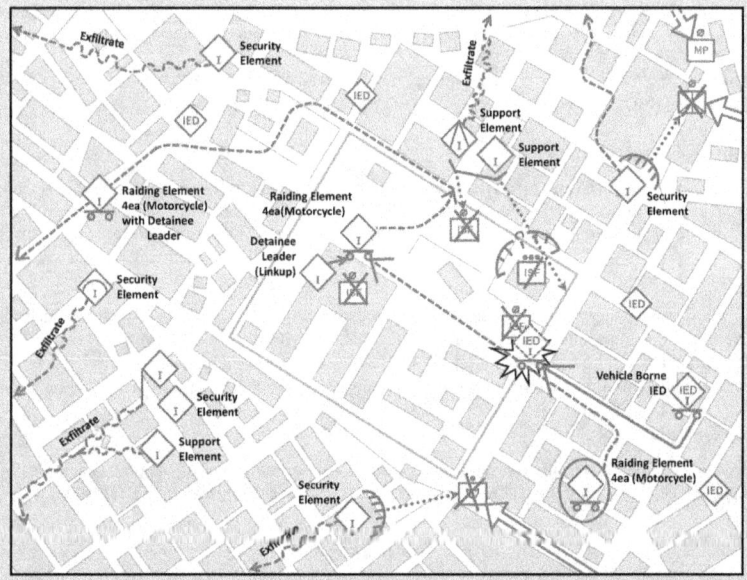

Figure 7-5. Insurgent raid (example)

A raiding element of four motorcycles moves through the breach at the main gate to a designated linkup point inside the facility that had been coordinated within a covert operative of the insurgency. The covert operative kills guards inside the holding cells area and releases the insurgent leader. Once the raiding element secures the insurgent leader, the element exits the facility through a side gate and quickly departs the area on motorcycles to a safe house.

During the raid, security elements positioned along likely avenues for enemy response forces ambush and delay law enforcement and quick reaction forces as they approach the detention facility. The local insurgent leader in command of the raid determines that the insurgent leader secured from the facility is now safe and directs support and security elements to exfiltrate from the area.

An INFOWAR cell video and audio records the raid from several vantage points. It releases an account of the successful raid on the Internet and to regional media outlets within hours of the raid.

(continued)

Functional Tactics

An INFOWAR cell video and audio records the raid from several vantage points. It releases an account of the successful raid on the Internet and to regional media outlets within hours of the raid.

Guerrilla Raid

A guerrilla battalion could employ a raid to demonstrate the inability of the enemy to effectively defend critical infrastructure. However, the battalion in this example has suffered significant losses, now having only one of its original three guerrilla companies. Therefore, it is reluctant to attempt a raid against an enemy POL installation by physically entering the objective, which has an enemy motorized infantry company and two combat outposts in the vicinity. However, the guerrilla battalion still has two of the original three 107-mm multiple rocket launchers (MRLs) of its weapons company, although it has lost the crews trained to operate them. So the battalion commander has one guerrilla platoon from his remaining company task-organized as an MRL platoon and moved to a designated site for specialized training in MRL tactical operations. SPF advisors from a neighboring state deploy into the area to train the task-organized platoon and continue plans and rehearsals for the MRL raid. Insurgent leaders and senior guerrilla commanders position themselves to observe the raid. An INFOWAR cell of the local insurgent organization positions to record the raid.

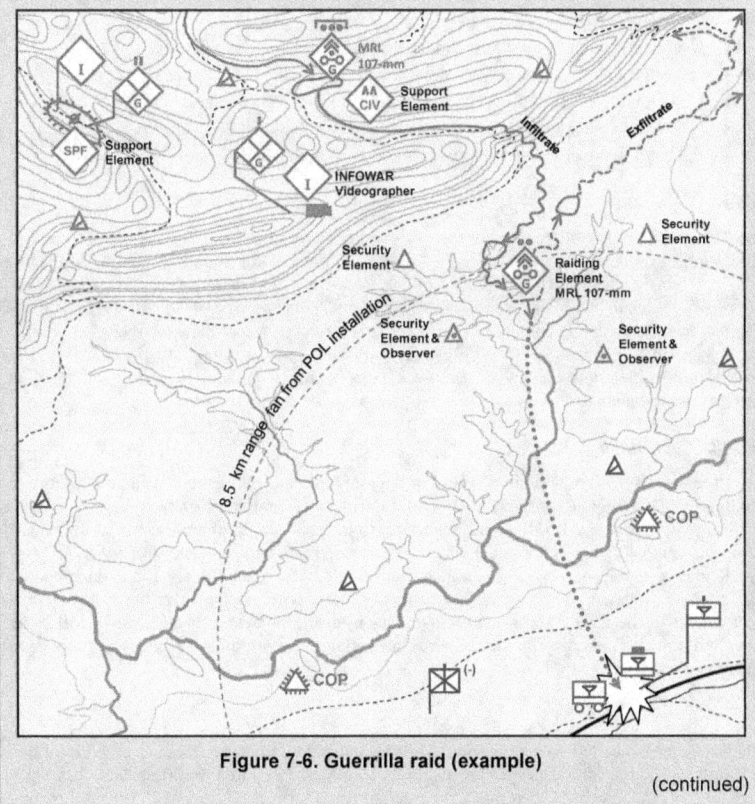

Figure 7-6. Guerrilla raid (example)

(continued)

> In this example, the raiding element is one MRL section from the task-organized MRL platoon, with one MRL (broken down into man-packable loads), the mortar crew, and additional riflemen (from the original guerrilla platoon). It infiltrates to a firing point near the maximum range of the rockets and establishes observations posts and security teams. With their weapon reassembled and ammunition ready for firing, the section leader orders the firing to commence. Significant damage is accomplished with three volleys of rockets from the MRL. The guerrillas quickly disassemble the weapon and move to hide positions along multiple exfiltration routes. The INFOWAR cell video and audio records the raid from several vantage points and releases an account of the successful raid on the Internet and to regional media outlets with a message from the local insurgent organization leader. On order, the guerrillas split into small teams and continue to exfiltrate to safe havens in the neighboring mountain range.

Fire Support

7-87. Most raids may use support elements in proximity to raiding elements with supporting direct fires from light, medium, or heavy machineguns and/or ATGLs. In some cases, support element(s) can also provide indirect fires from mortars or rockets allocated to the raiding force.

7-88. Fire support for a raid can assist in—
- Isolating a point of penetration in the enemy defenses at the objective.
- Fixing enemy forces in the objective and protecting the breach of enemy defenses.
- Suppressing effective direct and indirect fires from enemy forces.
- Disrupting enemy response forces.
- Obscuring the vision and sensors of the enemy as the raiding force withdrawals.

Air Defense

7-89. The irregular OPFOR recognizes that air defense is an all-arms effort that uses all weapon systems and resources available to the raid. (See TC 7-100.2, chapter 11 for more information on the OPFOR all-arms air defense concept.) The irregular OPFOR seeks new and adaptive ways to employ systems not traditionally associated with air defense. However, guerrilla battalions typically have a limited MANPADS capability in their weapons company. If allocated to a raiding force, these MANPADS would likely be located in security element(s). Security elements provide early warning and fires against enemy aerial response forces. (See note under Air Defense in Support of an Ambush regarding additional air defense assets that may be available.)

Engineer-Like Capabilities

7-90. Irregular OPFOR cells and units do not have organic mobility and/or countermobility capabilities. Breaching and/or removing obstacles before and during a raid on an objective may require insurgents or guerrillas with specialized skills and expertise from their civilian professions or previous military experience. Guerrilla sappers are not engineers but can perform some engineer-like functions. The irregular OPFOR may also rely on assistance from affiliated SPF or regular military forces in support of a raid. Tasks in either situation involve ensuring freedom of movement and maneuver to the objective, within the objective and to the target, and timely withdrawal from the objective. Insurgents, guerillas, and/or SPF soldiers with these skills may be located with the raiding, support, and/or security element depending on the assigned tasks.

Logistics

7-91. Raids are typically brief in duration. The raiding force will move from a secured location such as a safe house or safe haven to the objective area with all materiel required for the raid. Caches can be used in the vicinity of the objective when this technique improves tactical security and/or when irregular OPFOR materiel needs to be located close to the objective. Caches can also be mobile in transportation masked within a relevant civilian population in urban or rural environments. Examples of mobile caches are as simple as a wheel barrow covered with a tarpaulin to a modern truck with hidden compartments as part of a commercial convoy.

INFOWAR

7-92. INFOWAR primarily supports a raid by concealing the action through deception and information protection. Successful raids can be used in INFOWAR to demonstrate the inability of enemy forces to do one or more of the following:
- Defend civilian, government, and military facilities.
- Safeguard key representatives of the governing authority.
- Protect civilian and military infrastructure critical to a governing authority's counterinsurgency or counter-guerrilla campaign.

(See appendix A for more information on irregular OPFOR INFOWAR capabilities.)

7-93. Recurring irregular OPFOR raids can be conducted to—
- Isolate psychologically the enemy force and governing authority from each other and the relevant civilian population.
- Degrade the resolve of the enemy force and governing authority to continue counterinsurgency or counter-guerrilla operations.
- Obtain needed materiel to continue the irregular OPFOR.

RECONNAISSANCE ATTACK (GUERILLAS)

7-94. A *reconnaissance attack* is a tactical offensive action that locates moving, dispersed, or concealed enemy elements and either fixes or destroys them. A guerrilla commander may also use it to gain information about the enemy's location, dispositions, military capabilities, and possibly his intentions. (Insurgent cells typically do not have sufficient combat power to conduct a reconnaissance attack.)

7-95. The guerrillas recognize that an enemy will conduct significant measures to prevent them from gaining critical intelligence. Therefore, quite often they will have to fight for information, using an offensive action to penetrate or circumvent the enemy's security forces to determine who and/or what is located where or doing what.

7-96. The reconnaissance attack is the most ambitious method to collect information and is ordered by a guerrilla commander only after prudent consideration of other tactical alternatives. Key factors in reconnaissance attack considerations are—
- Accurate situational assessment of an area of responsibility (AOR) within which the guerrillas operate.
- Contact conditions: having maintained contact with the enemy or the requirement to reestablish contact.
- Current capabilities to simultaneously support the movement and/or maneuver of multiple reconnaissance, security, and action elements in the guerrilla AOR.
- Level of active support from the local relevant population.

Functional Organization for a Reconnaissance Attack

7-97. Depending on the situation, a guerrilla commander organizing a reconnaissance attack may designate reconnaissance, security, and/or action elements. There may be more than one of each type. The commander may also form various types of support elements.

Reconnaissance Element(s)

7-98. If the purpose of the reconnaissance attack is merely to gain information, a guerrilla commander may organize several reconnaissance elements. Their role is to locate enemy elements operating in the unit's AOR. If the purpose is to also have the capability to fix and/or destroy located enemy elements, the reconnaissance elements provide reconnaissance and surveillance support to the elements that carry out

Chapter 7

those functions. A tactical option is for security elements to perform this role if reconnaissance elements are not formed. Once a reconnaissance element locates an enemy element, it may become a security element.

Security Element(s)

7-99. If the guerrilla commander believes he has sufficient combat power to engage enemy elements that may be located, he may also organize one or more security elements. Size and task organization of security elements are dependent on the assigned mission and the expectation of how long a security element will conduct its functional tasks before arrival of action elements or other elements. (When performing some of these functional tasks, the unit originally designated as a security element may receive a designation that describes that specific function.)

7-100. Security elements can either work in conjunction with reconnaissance elements or perform the reconnaissance role for the guerrilla unit. Upon locating an enemy element, a security element may be directed to conduct one of several tactical tasks. It may—

- Report on conditions in the AOR.
- Observe and monitor the enemy in the AOR.
- Locate enemy direct-fire weapons concentrations such as enemy antitank ambush sites.
- Locate enemy countermobility obstacles along friendly axes.
- Identify bypasses to enemy countermobility obstacles.
- Prepare infiltration lanes through enemy countermobility obstacles that cannot be bypassed.
- Fix enemy forces (as a *fixing element*).
- Block probable enemy avenues of withdrawal or reinforcement (as a *blocking element*).
- Attack a smaller enemy element (as an *ambush element* or *raiding element*).

Action Element(s)

7-101. A guerrilla commander task-organizes one or more action elements to conduct designated tasks against enemy element(s). These action elements may receive a functional designation that more specifically describes the action they are to accomplish, such as *raiding element*. Once an enemy element is located and/or fixed, the action element(s) attack to defeat or destroy the enemy. The number of action elements for a mission is based on the anticipated tasks to engage enemy elements located by reconnaissance and/or security elements. Action elements may be directed to plan for operations in conjunction with one or more security elements in an AOR. Each security element may not necessarily have an action element dedicated to follow and assume an offensive task.

Support Element(s)

7-102. One or more support elements can perform various supporting tasks. Typical CS and/or CSS tasks are discussed in subsequent paragraphs under Support of a Reconnaissance Attack.

Executing a Reconnaissance Attack

7-103. Multiple attack axes often characterize reconnaissance attacks. Control measures may include start times, check points, orientation objectives, and objective rally points designated for each axis. Multiple axes of advance provide more tactical flexibility to guerrilla elements in accomplishing their respective missions and tasks (see figure 7-7 on page 7-21).

7-104. Guerrilla elements normally infiltrate within or into an AOR. The norm is reconnaissance and/or security elements maneuvering separately to find the enemy and/or report on activity and/or conditions along designated routes or axes. When the enemy is located, the guerrilla commander decides on subsequent tasks for his unit. A typical task is to fix the enemy with security forces and attack with action elements to defeat or destroy the enemy. Success often relies on the ability of reconnaissance and/or security elements to operate independently. Action elements are positioned within an AOR to quickly respond to intelligence confirmed by reconnaissance and/or security elements. (See figure 5-7 for an example of a reconnaissance attack.)

Functional Tactics

Reconnaissance Attack Example

Recent combat actions have severely reduced a guerrilla battalion two guerrilla companies and limited organic fire support. The guerrilla brigade allocates mortar and rocket launcher support as the guerrilla battalion prepares for a reconnaissance attack. The battalion commander uses remnants of his third company to create two dummy companies as deception elements. SPF INFOWAR teams support guerrilla deception activities. Avoiding enemy screening elements, several reconnaissance elements infiltrate to reestablish contact with the enemy. Reconnaissance elements report on activities and conditions within the AOR, identify enemy force locations, and guide security elements into positions to fix enemy forces. Other reconnaissance elements reach their orientation objectives without enemy contact and transition to security tasks. The guerrilla battalion conducts simultaneous attacks on enemy forces using massed direct and indirect fires.

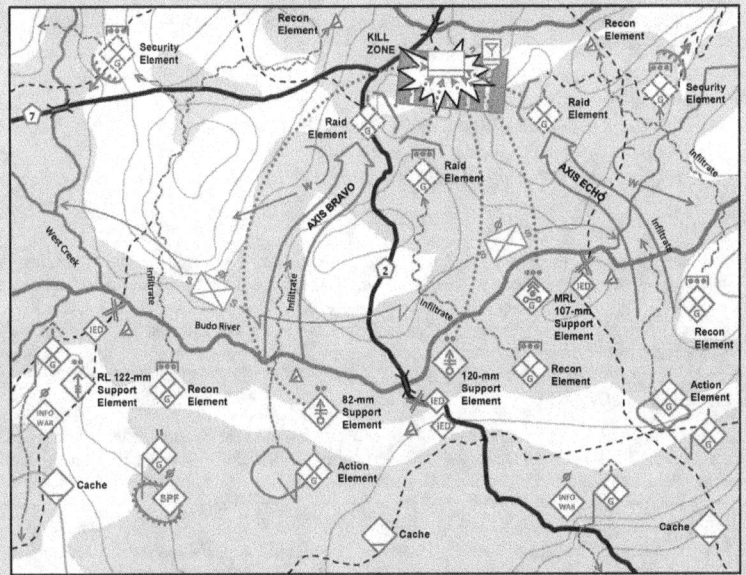

Figure 7-7. Guerrilla reconnaissance attack (example)

Guerrilla forces quickly withdraw to safe havens while security elements and on-call indirect fires prevent any enemy pursuit. The surprise attack by guerrillas disrupts enemy logistics and stalls enemy offensive operations.

Command and Control of a Reconnaisance Attack

7-105. A reconnaissance attack requires planning and coordination typically more detailed than in other offensive actions for a guerrilla unit. The commander or leader must plan for multiple reconnaissance and/or security elements operating across a broad area is one consideration. He must also position one or more action elements to respond quickly to tactical opportunities identified by reconnaissance and/or security elements. The organization and positioning of functional elements is further complicated as guerrillas plan to deceive the enemy as to their presence and tactical intentions.

Support of a Reconnaissance Attack

7-106. A reconnaissance attack typically requires several types of support. These can include reconnaissance, fire support, air defense, engineer-like capabilities, logistics, and INFOWAR.

7-107. Special situations may exist when guerrilla units have levels of capability more expected of regular forces. These capabilities can include armored fighting vehicles, aviation, and additional air defense or fire support. Weapons and equipment can be—
- Provided by state and/or non-state sponsors.
- Acquired from indigenous resources.
- Captured from the enemy.
- Purchased from criminal organizations and/or other commercial enterprise intermediaries.

7-108. Sustained intelligence activities, logistics, and training support are critical aspects of long-term guerrilla unit effectiveness. These types of functional capabilities of a guerrilla unit can be supported by—
- SPF from a state overtly or covertly affiliated with the guerrilla unit.
- Regular military force or intelligence activity liaison teams that operate with the guerrilla unit.
- Criminal organizations with expertise and assets available on a contractual basis to the guerrilla unit.
- Adherents in the relevant population who actively support guerrilla actions but will not be visibly involved in the direct actions and combat of the guerrilla unit.
- Higher-level or SPF-supported INFOWAR activities.

Reconnaissance

7-109. Reconnaissance in a reconnaissance attack can be conducted in two primary ways. A reconnaissance element can be formed with a mission to find the enemy forces and guide security elements to locations that allow the security elements to fix the enemy forces. When no reconnaissance element is formed, each security element performs its own reconnaissance tasks with the responsibility to both find and fix an enemy force.

Armored Fighting Vehicles

7-110. Some guerrilla units may have armored vehicles present in their organization and can use them for particular tasks. When a guerrilla commander believes he has a level of near parity with local enemy forces and is confident in his air defense capabilities for limited periods of time, he can use armored vehicles in his various functional element(s). With effective camouflage, concealment, cover, and deception (C3D), he may decide to introduce wheeled and/or tracked armored vehicles from safe havens or local hide positions to augment reconnaissance and/or security elements. The thermal imagers and other electro-optical aids on armored vehicles can be of great value in detecting enemy forces. Armored vehicles can also increase mobility, firepower, and protection for reconnaissance and/or security elements, significantly enhancing their ability to fix and possibly destroy the enemy. The mobility and speed of these vehicles permit them to serve in the action element(s), rapidly orienting on located and/or fixed enemy forces and moving to a position of advantage to destroy them. Especially when the enemy was previously unaware of their presence in the AOR, these armored vehicles can provide a significant advantage in firepower and shock effect.

Fire Support

7-111. Fire support is positioned in an AOR to provide responsive fires throughout the reconnaissance phase, security element movement and occupation of fixing and/or blocking positions, and the maneuver and combat of action of elements. Various wheeled, towed, or tracked fire support systems may be available to cover the withdrawal of reconnaissance, security, action, or other support elements after completing the reconnaissance attack mission.

7-112. Guerrilla units may include medium and heavy mortars, artillery, and/or rocket systems as a varied collection of antiquated and contemporary capabilities. Mobility of systems can vary also. Guerrilla units may have to improvise transportation, using captured military trucks, local commercial vehicles that conceal a weapon system, or other expedient means. In some cases, pack animals, bicycles with reinforced frames, and/or porters can be used to transport disassembled fire support systems for reassembly at locations unexpected by the enemy.

Aviation

7-113. Aviation is not typically anticipated in a guerrilla unit order of battle. However, a guerrilla commander can acquire small unmanned aerial vehicles (UAVs) with relative ease through normal means on the commercial market and/or illicit purchases from criminal organizations or other sources. State sponsors can provide technical expertise with clandestine agents and/or technicians or provide training and advice through SPF teams. A guerrilla commander could acquire systems and technical expertise on a contractual basis from disenfranchised experts or rogue non-state enterprises. UAVs in a guerrilla unit can be used to enhance real-time reconnaissance and surveillance in conjunction with reconnaissance elements maneuvering on the ground in an AOR. A number of small UAVs could be armed with IEDs to create an aerial attack capability. Used with a swarm technique on a point or area target, a mass of low-flying UAVs could conduct a disruptive or devastating attack on critical enemy assets.

7-114. Aviation in guerrilla units can also include rotary- and/or fixed-wing capabilities. Although these capabilities may be considered exceptional for most guerrilla forces, state and non-state sponsors can covertly introduce limited aviation capabilities into an advanced insurgency and/or highly trained guerrilla units. Operating from a guerrilla enclave or safe haven, even one or two helicopters or fixed-wing commercial aircraft could be converted into an attack system with military-grade rockets or bombs. Capability for aviation attack against a governing authority or its military forces could create a significant psychological impact in a guerrilla INFOWAR campaign.

Air Defense

7-115. Guerrilla air defense can be developed to a level of protection that limits enemy aerial response forces and reinforcements from influencing a particular guerrilla mission in the AOR. Any guerrilla unit conducts all-arms air defense as a norm to damage and/or destroy tactical enemy aircraft within the range of their available small arms weapons systems. (For more information on the all-arms air defense concept, see chapter 11 of TC 7-100.2.) However, guerrilla battalions also typically have a limited MANPADS capability in their weapons company. (See note under Air Defense in Support of an Ambush regarding additional air defense assets that may be available.)

Engineer-Like Capabilities

7-116. Mobility and countermobility support to a reconnaissance attack focuses on freedom of movement and/or maneuver of reconnaissance, security, and action elements. Mobility and countermobility tasks are performed by guerrillas with specialized skills. For example, guerrillas with expertise from civilian engineering occupations and/or previous training by SPF teams may be assigned tasks that concentrate on emplacing rudimentary obstacles and IEDs along planned withdrawal routes to disrupt any pursuit by the enemy after a successful reconnaissance attack. Guerrillas from sapper units can be task-organized to assist various elements in infiltrating through enemy security elements, breaching enemy obstacles, as well as support attacks on located enemy forces or installations.

7-117. Additional mobility and countermobility support, training, and assistance can be obtained from sappers in SPF teams that accompany and/or augment a guerrilla unit. SPF and/or guerrilla sappers can also train active supporters in the relevant local population to assist the reconnaissance attack in specified supporting roles. Guerrilla units can use to their advantage the blurred distinctions of what constitutes the role of active support versus being considered a guerrilla. Regardless of who is providing mobility and countermobility support, this guerrilla capability is essential to offensive action in fixing, blocking, and/or attacking enemy forces.

> *Note.* Guerrilla sappers are *not* combat engineers. However, sappers are trained to perform several tasks that are typical of raider and combat engineer-like functions.

Logistics

7-118. A reconnaissance attack can be conducted by widely dispersed guerrilla elements operating over extended time periods and distances. Elements typically carry sufficient logistics with them during their movements. Guerrillas can create and stock caches and/or preposition designated logistics elements in an AOR for multiple reconnaissance advances toward orientation objectives and during guerrilla unit withdrawals to safe havens after a reconnaissance attack.

7-119. Guerrillas are self-sufficient as a goal but must often subsist on the local economy without offending a local relevant population. Use of civilian facilities and support of the population to guerrilla units may result in reaction by enemy forces and the governing authority on the civilian population and its institutions. Guerrilla INFOWAR activities amplify enemy repression to increase and sustain tactical and logistics support by a relevant population for guerrilla actions.

INFOWAR

7-120. With the support of INFOWAR deception, the guerrilla commander attempts to deceive the enemy concerning the strength and composition of his forces, their current deployment and orientation, and the intended manner of employment. False intelligence provided to enemy forces by active supporters of the guerrillas disrupts enemy information collection. When successfully conducted, deception activities support tactical surprise by the guerrilla force and improve the likelihood of achieving the reconnaissance attack objective.

7-121. INFOWAR activities in a reconnaissance attack are primarily executed to—
- Deceive the enemy force regarding guerrilla actions and intentions.
- Protect elements of the guerrilla force from being detected.
- Create a false sense of security in the enemy.
- Encourage the active and passive support of the guerrilla operations by a relevant population.
- Demoralize enemy forces and the governing authority they support.

TYPES OF DEFENSIVE ACTION

7-122. Insurgents and guerrillas can employ some of the types of defensive action also used by smaller tactical units of the regular OPFOR. Such actions can include—
- Defense of a simple battle position.
- Defense of a complex battle position.

(See TC 7-100.2 for basic discussion of these types of defensive actions, as they are also conducted by the regular OPFOR.)

7-123. Irregular OPFOR leaders and commanders select the defensive action best suited to accomplishing their mission, given the conditions under which they assume a defensive posture. Some parts of an insurgent or guerrilla organization may conduct defensive actions while other parts of the same organization are on the offense.

BATTLE POSITIONS

7-124. A battle position (BP) normally is a defensive location oriented on a likely enemy avenue of approach. However, the irregular OPFOR may select defensive locations to avoid contact with an enemy but provide for defense if discovered. When irregular OPFOR leaders determine that they will operate in a defensive posture, defensive positions will be either a simple battle position (SBP) or complex battle position (CBP). The mission and specific circumstances will influence the type of BP to establish and occupy. Figure 5-8 shows examples of symbols for SBPs and CBPs (see figure 7-8 o page 7-25.

Functional Tactics

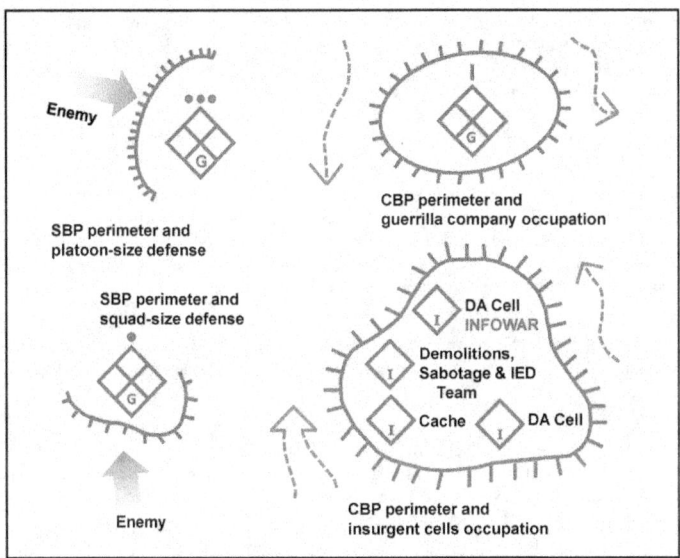

Figure 7-8. Simple and complex battle positions

Note. Sometimes graphics show a relatively large unit, such as a guerrilla battalion or brigade, inside a symbol for a CBP. This actually means that such a unit's subordinates occupy a series of CBPs within that area.

Simple Battle Position

7-125. A *simple battle position* (SBP) is a defensive location oriented on the most likely enemy avenue of approach. SBPs are not necessarily tied to complex terrain. However, they often employ as much fortification and C3D measures as time allows. Defenses are improved upon continuously until the SBP is abandoned.

Complex Battle Position

7-126. A *complex battle position* (CBP) is a defensive location designed to employ a combination of complex terrain, C3D, and engineer-like capabilities to protect the cells or units within them from detection and attack while denying their seizure and occupation by the enemy. CBPs typically have the following characteristics that distinguish them from SBPs:

- *Not* on or along an enemy avenue of approach.
- Limited avenues of approach toward and/or in vicinity of a CBP.
- Observation of any existing avenues of approach.
- Defensive posture with an integrated 360-degree perimeter.
- Countermobility and mobility efforts prioritizing C3D measures of the CBP location.
- Substantial logistics caches.
- Sanctuary.

DEFENSE OF A SIMPLE BATTLE POSITION

7-127. Construction of an SBP places special attention on the camouflage, concealment, and cover of fighting positions in urban and rural terrain. The irregular OPFOR normally expects significant enemy reconnaissance, intelligence, surveillance, and target acquisition (RISTA) capabilities and recognizes that sophisticated RISTA capabilities may be supporting the enemy. An effective counter to such levels of sophisticated technology and systems may be to embed the SBP within a relevant population in an urban and/or rural environment, or physically use rural and/or urban terrain to mask the presence of SBPs. Examples include the use manmade underground shelters, tunnels, natural shelters such as caves, and/or village or city dwellings. An SBP or group of SBPs establishes kill zone(s) on likely enemy avenues of approach.

7-128. Deceptive techniques can include the façade of being commercial or private equipment, vehicles, work places, and/or public institutions and public gathering places such as houses of worship, hospitals, and civic centers with regular intermingling of the relevant population. Insurgents usually wear the clothing of the local population and often keep weapons, munitions, and materiel in caches that are easily retrievable in the vicinity of the SBP. The same may be true of guerrillas. However, guerrillas may transition to recognizable paramilitary uniforms.

7-129. The irregular OPFOR commander or leader makes prudent risk assessments when establishing SBPs. He evaluates the desirability and/or requirement to invest substantial time, effort, and materiel on an SBP. He weighs this against the expectation that he must defeat an enemy that can typically mass combat power quickly against an SBP.

7-130. Once the commander or leader decides to defend an SBP, he focuses his available combat power on one or more kill zones. The irregular OPFOR plans and rehearses all actions necessary to prevent enemy penetration of an SBP and/or what an SBP or group of SBPs is protecting, and also considers measures to defeat an enemy penetration of an SBP if it occurs.

7-131. The commander or leader considers what criteria he will use to direct a withdrawal and/or withdrawal under pressure from an SBP or group of SBPs. Unless directed to retain a specific SBP by a higher level, the commander or leader responsible for an SBP recognizes that he is committed to a long-term struggle and that preserving combat power for a future engagement may be the appropriate decision. However, some insurgents or guerrillas may have a self-determined commitment or directed mission to fight until killed or captured in a particular SBP.

Functional Organization for Defending an SBP

7-132. The commander or leader defends an SBP with cells or units that are organized as functional elements. Typical functional designations are—
- Disruption element.
- Main defense element.
- Reserve element.
- Support element.
- Deception element.

7-133. There may be more than one of each type. The name of an element describes its function within the defensive action.

Disruption Element(s)

7-134. Insurgents or guerrillas assigned to a disruption element have a mission of identifying enemy reconnaissance efforts and reporting the location, disposition, and composition of approaching enemy forces. When disruption elements have the capability to target and attack designated subsystems of an enemy force, they conduct disruption actions as part of a comprehensive defense plan of the higher commander or leader.

7-135. Disruption activities may include direct and indirect fires, remote-controlled or command-detonated IEDs and/or other execution of obstacles to slow, channel, contain, or block an enemy force. The normal intention of a disruption element is to not become decisively engaged by the enemy. However, a commander or leader can direct decisive engagement if the action is necessary to preserve the combat power of other critical capabilities in the irregular OPFOR organization.

7-136. Tactical tasks typical of a disruption element include—
- Ambush.
- Attack by fire.
- Delay.
- Disrupt.

7-137. The irregular OPFOR will typically not assign a small cell or unit a fixing task when an expectation of "fix" is to deny movement of any part of an enemy force. A more probable task for the irregular OPFOR in an SBP is "delay" with an expectation to slow the momentum of an enemy advance and cause significant damage to the enemy force without becoming decisively engaged.

7-138. A disruption element for an SBP can be as small as one or two insurgents or guerrillas with assault rifles, light and/or medium machineguns, grenade launchers, IEDs, and/or ATGLs. Typically, it is no larger than 8 to12 such personnel.

Main Defense Element(s)

7-139. The main defense element of an SBP is responsible for defeating an attacking force. Insurgents or guerrillas in a main defense element are prepared to use fires and maneuver to defeat the penetration or seizure of their SBP or other SBPs. Main defense elements focus the combat power of available weapon systems into designated kill zones to defeat or destroy an enemy force.

Reserve Element(s)

7-140. The reserve element of an SBP exists to provide the irregular OPFOR commander or leader with tactical flexibility. The commander or leader will normally assign priorities of effort to the reserve element for contingency planning and rehearsals. Although a reserve element may not have specified tasks to perform initially, tactical tasks it can later receive include—
- Counterattack.
- Block.
- Contain.
- Delay.
- Defend.

Support Element(s)

7-141. The support element of an SBP can include—
- CSS.
- C2.
- Direct fires such as heavy machineguns, ATGLs, antitank guided missiles, recoilless rifles, or automatic grenade launchers.
- Indirect fires such as mortars or rockets.
- Nonlethal actions such as smoke obscurants.
- INFOWAR activities.
- Engineer-like capabilities with specialized talents of individual insurgents or guerrillas.

Deception Element(s)

7-142. To keep the enemy from discovering the nature of the defenses and to draw fire away from actual elements, the defending force may establish dummy firing positions and battle positions. In addition to enhancing force protection, the irregular OPFOR may use these deception elements as an economy-of-force measure to portray strength where none exists.

Chapter 7

Organizing the Battlefield for an SBP

7-143. When establishing an SBP, the commander or leader of the defending force specifies the organization of the battlefield from the perspective of his level of command. This normally includes a battle zone and often a support zone. It may also include a disruption zone. (See figures 7-9 and 7-10 for examples of SBP defense.)

Battle Zone

7-144. The battle zone is the area where the defending commander or leader commits the preponderance of his force to the task of defeating attacking enemy forces. Generally, an SBP will have its battle zone fires integrated with those of any adjacent SBPs. Fires will orient to form kill zones where the defenders plan to destroy key enemy targets.

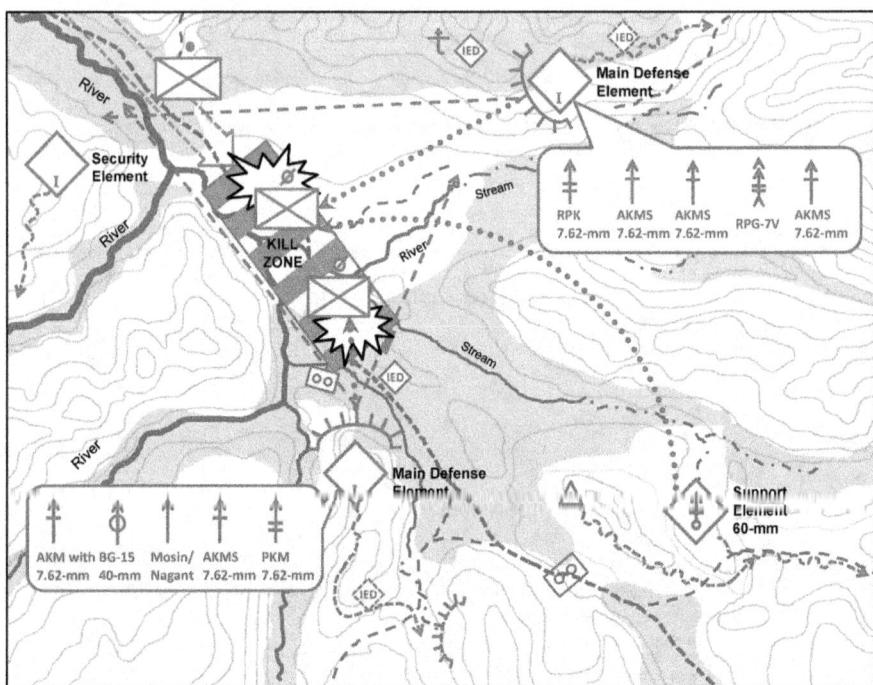

Figure 7-9. Insurgent defense of a simple battle position (example)

Disruption Zone

7-145. The disruption zone is the area outside the battle zone where the defenders may seek to—
- Report on the enemy situation.
- Defeat enemy reconnaissance efforts.
- Detect attacking forces.
- Disrupt and delay an attacker's approach.
- Engage and destroy key attacking elements prior to engagement in the battle zone.

7-146. Engagement (such as ambush or attack by fire) in the disruption zone may be beyond the capabilities of a small disruption element. However, it may be able to delay or disrupt the enemy advance or channel it away from the SBP(s). In any case, some level of reconnaissance and/or security will be placed outside of an SBP for early warning of enemy approach.

Support Zone

7-147. Depending on the mission and size of the defensive positions, support capabilities may be incorporated into the battle zone of an SBP, or there may be a support zone inside or outside the battle zone. Aside from the support element(s), the support zone may contain a reserve element.

Executing Defense of an SBP

7-148. Aggressive security measures throughout the development and occupation of an SBP provide early warning of enemy activities. Once enemy forces are detected, the irregular OPFOR commander or leader decides when to engage the enemy. He may direct that disruption elements engage with direct and indirect fires, or he can continue to observe movements and maneuver of the enemy as it approaches SBP kill zones.

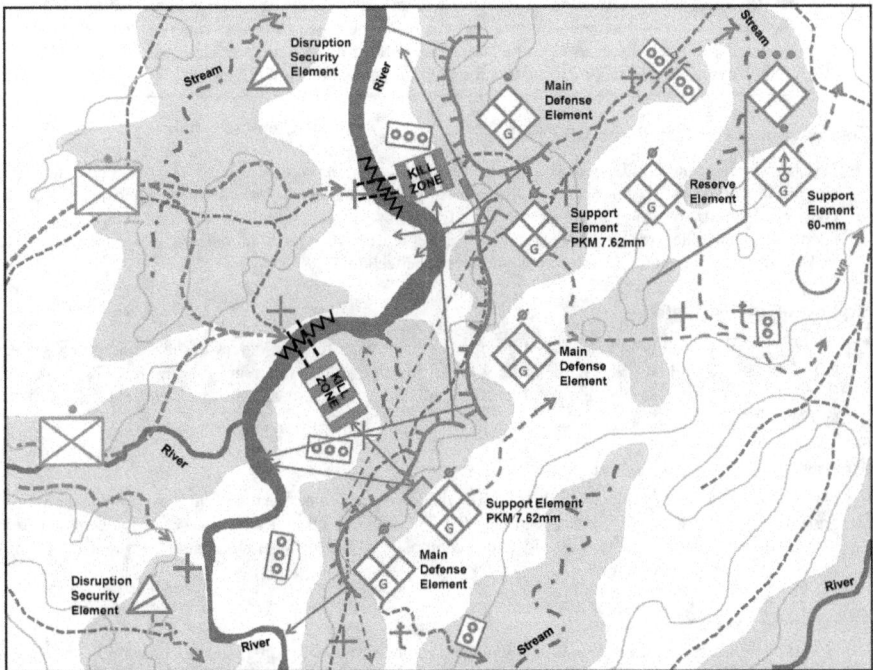

Figure 7-10. Guerrilla defense of a simple battle position (example)

7-149. On order of the commander or leader, the irregular OPFOR engages the enemy to defeat and/or destroy the enemy in designated kill zones. Given the norm of an enemy quickly responding to contact with additional forces, the commander or leader may decide to withdraw or reposition from initial defensive positions.

7-150. The irregular OPFOR can initiate deception activities to confuse an enemy when it is displacing from its initial SBP or SBPs. Deception can include small stay-behind direct action cells or guerrilla teams to indicate a continued defense of an SBP or SBPs. Other actions that can slow the reaction of enemy forces to irregular OPFOR maneuver are keeping insurgents or guerrillas close to or within a relevant civilian population as a shield. Multiple IEDs along withdrawal routes can be both active and inert munitions. Both types of IED-appearing obstacles cause enemy responses to assess and disarm or bypass the munitions.

> *Note.* The commander or leader of an SBP can position main defense elements to mass overlapping small arms fire (SAF) and ATGL fires in to a kill zone or kill zones. He selects the kill zone(s) to best use restrictions in terrain, such as difficult fords across a river. He reinforces the terrain with antipersonnel mines to further channel, separate, or contain the enemy force as it enters the kill zone(s). Security elements are located where they can observe any approaching enemy along ground and aerial avenues of approach to the SBP. The security elements may be directed to continue reporting on enemy activity and to not engage the enemy. Support elements such as mortars are ordered to commence indirect fires in support of machinegun and automatic rifle SAF that coincide with detonation of antipersonnel mines and/or IEDs in and near the kill zone. All SAF orient on identified enemy in their kill zone(s).
>
> When the commander or leader determines that he has achieved the intended effects on the enemy, he directs exfiltration of his elements and rendezvous at a designated location. Main defense elements disengage while support elements continue to suppress enemy forces. Security elements may add their SAF to the suppression. Security and support elements disengage but may leave observers in the area to continue reporting on actions of the enemy.

Command and Control of an SBP Defense

7-151. The irregular OPFOR commander or leader will position himself where he can best command and control the defensive fight. Security during defensive preparations includes communications such as couriers, landline or wire intercommunication systems, visual signals, and limited use of cellular telephones. Once the irregular OPFOR initiates the defense with direct and indirect fires, communications make full use of handheld radios and cellular telephone technology.

Support of an SBP Defense

7-152. Support of an SBP defense is typically provided from local resources in the geographic area and may include CS and/or CSS. While some of this support resides in irregular OPFOR cells or units, a higher-level organization may temporarily allocated some support or assign it to a particular subordinate cell or unit.

Reconnaissance

7-153. SBP defenders will perform aggressive counterreconnaissance activities to prevent the enemy from remaining in reconnaissance contact with the SBP. The irregular OPFOR will observe avenues of approach to provide early warning; determine location, composition, and disposition of attackers; and direct fires against key enemy capabilities.

Armored Fighting Vehicles

7-154. The irregular OPFOR may occasionally capture armored fighting vehicles or acquire them from other sources operating in the region. When possessed by the irregular OPFOR, armored fighting vehicles are normally concealed and covered in hide positions until the commander or leader directs them into the engagement. Their visible and other signatures are masked to preclude identification by enemy RISTA systems. Once armored fighting vehicles are employed in defensive actions, they move frequently to alternate and supplemental fighting positions to improve their survivability.

Fire Support

7-155. Fire support to an SBP or SBPs is usually under the command and control of the commander or leader responsible for the defensive actions. When additional fire support assets are required in a defensive mission, a higher-level organization may allocate assets to a subordinate for a particular mission or on a temporary basis.

Air Defense

7-156. The irregular OPFOR in an SBP can employ active and passive air defense measures to protect the defender from air threats within an all-arms air defense concept. Medium or heavy machineguns and shoulder-fired MANPADS may be found in or near an SBP. (See note under Air Defense in Support of an Ambush regarding additional air defense assets that may be available.)

Engineer-Like Capabilities

7-157. The irregular OPFOR commander or leader is responsible to countermobility and mobility tasks. He uses the specialized talents that exist among insurgents or guerrillas in a cell or unit but does not have cells or units structured for only engineer-like tasks.

Logistics

7-158. When present, logistics capabilities will normally be found with the support element, within the SBP. However, they can also be located in caches and safe houses in or near an SBP. Weapons, munitions, and materiel will normally be hidden from open surveillance until just prior to a defensive action. Items are brought from concealment and emplaced in fighting positions on order of the irregular OPFOR commander or leader.

INFOWAR

7-159. The SBP obtains support from INFOWAR activities that deceive the enemy as to the defenders' presence, actual locations, and/or intentions. A successful INFOWAR campaign can promote the impression that the enemy to failing to effectively protect its population. Other INFOWAR messages portray the enemy force and governing authority as a corrupt regime and further isolate them from a relevant population the irregular OPFOR claims to protect. (See appendix A for additional information on irregular OPFOR INFOWAR capabilities.)

DEFENSE OF A COMPLEX BATTLE POSITION

7-160. C3D measures are critical to the success of a CBP, since the defenders generally want to avoid enemy contact. Additionally, cells or units in a CBP will remain dispersed to negate the effects of precision ordnance strikes. Generally, once the defense is established, non-combat vehicles will be moved away from concentrations of personnel to reduce their signature on the battlefield.

7-161. Units defending in CBPs will use restrictive terrain and countermobility efforts to deny the enemy the ability to easily approach the position. Construction of a CBP places special attention on the camouflage, concealment, and cover of fighting positions in urban and rural terrain. The irregular OPFOR normally expects enemy RISTA capabilities to be significant and recognizes that sophisticated RISTA capabilities may be supporting the enemy. An effective counter to such levels of sophisticated technology and systems may be to embed a CBP within a relevant population in an urban and/or rural environment. Examples include the use manmade underground shelters, tunnels, natural shelters such as caves, and/or village or city dwellings.

7-162. Cultural shielding is a tactical consideration to deny the enemy the ability to detect and attack a CBP. Examples of cultural shielding in order to create tactical standoff are using a religious location, school, community center, or medical facility as a base of fire or firing from within a crowd of noncombatants.

7-163. If a CBP is identified and attacked, the commander or leader will engage only as long as he perceives an ability to defeat the enemy. Prior to becoming decisively overmatched, he will withdraw in order to preserve his combat power. A guerrilla commander or leader can be directed by a higher guerrilla headquarters to accept decisive engagement in order to support a larger mission.

Functional Organization of a CBP Defense

7-164. The commander or leader of the defending force organizes his subordinates as functional elements. Typical functional designations are—
- Disruption element.
- Main defense element.
- Reserve element.
- Support element.
- Deception element.

7-165. These functional elements conduct tactical actions very similar to those used in defending an SBP. That is because the names of these elements identify their basic functions. However, the following paragraphs highlight some differences. There may be more than one of each type of element.

Disruption Element(s)

7-166. The disruption element of a CBP is primarily concerned detecting attackers and providing early warning to the defending force. The disruption elements may be directed to only observe and report enemy movements and maneuver but can also be directed to attack enemy forces once they pass the disruption elements.

Main Defense Element(s)

7-167. The main defense element of a CBP is responsible for defeating an attacking force. This element can be directed to delay an enemy while other cells or units withdraw from direct contact with the enemy.

Reserve Element(s)

7-168. The reserve element of a CBP exists to provide the commander or leader with tactical flexibility. Tasks for a reserve element can include—
- Counterattack.
- Block.
- Delay.
- Defend.

Support Element(s)

7-169. The support element of a CBP has tasks that include C2, CS, and CSS for the defending force. Other support functions can include direct and direct fires; countermobility or mobility capabilities; and/or INFOWAR activities. Support elements typically are located within the CBP but can be outside and in the vicinity of a CBP.

Deception Element(s)

7-170. To keep the enemy from discovering the nature of the defenses and to draw fire away from actual elements, the defending force may establish dummy firing positions and battle positions. In addition to enhancing force protection, the irregular OPFOR may use these deception elements as an economy-of-force measure to portray strength where none exists.

Organizing the Battlefield for a CBP

7-171. When establishing a CBP, the commander or leader of the defending force specifies the organization of the battlefield from the perspective of his level of command. He will determine if he designates a battle zone supported by a disruption zone and support zone. (See figures 7-11 on page 7-33 and 7-12 on page 7-34 for examples of a CBP defense.)

Functional Tactics

Battle Zone

7-172. The battle zone is the area where the defending commander commits the preponderance of his force to the task of defeating attacking enemy forces or delaying them while the defenders withdraw. In the defense of a CBP, the battle zone is typically the area in and surrounding the CBP that the defending force can influence with its direct fires. It may be larger depending on the availability of indirect fires.

Disruption Zone

7-173. The disruption zone is the area outside the battle zone where the defenders may seek to—
- Report on the enemy situation.
- Defeat enemy reconnaissance efforts.
- Detect attacking forces.
- Disrupt and delay an attacker's approach.
- Destroy key attacking elements prior to engagement in the battle zone.

Support Zone

7-174. The support zone contains C2, CSS, fire support, and other supporting assets. A reserve element may also be located there. The support zone is normally located within the CBP.

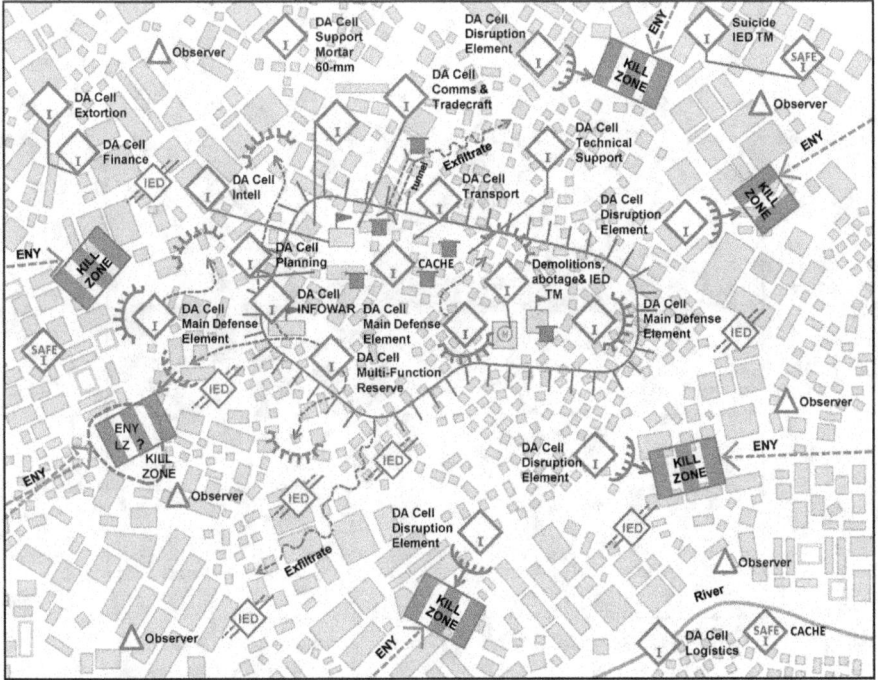

Figure 7-11. Insurgent defense of a complex battle position (example)

Executing Defense of a CBP

7-175. The commander or leader will determine whether or not to organize a disruption force in a disruption zone. He may determine that the manning and capabilities of his organization are more effectively used with security elements close to the defensive perimeter of the CBP. Whether near or distant from the CBP main defenses, security actions are disruption, active reconnaissance, and counterreconnaissance. However, contact with the enemy is on order of the commander or leader. The normal intention is to keep a CBP undiscovered by enemy forces. Working in close coordination with main defense elements, support capabilities are often incorporated into the CBP.

7-176. A disruption element in the disruption zone can engage enemy forces in tactical depth as they approach the CBP or CBPs. Direct and indirect fires coordinated by a disruption element can delay and attrit enemy forces and cause them to enter the attack on a CBP in a piecemeal manner.

7-177. Main defense elements mass direct and indirect fires to defeat an enemy attack. The commander or leader of the defending force may retain a reserve element and commit it only when necessary to prevent defeat by enemy forces. Support elements in the support zone provide support to defenders in the disruption zone and battle zone as required. In the event the commander or leader orders of a withdrawal from the CBP, some support elements will exfiltrate quickly while other elements such as indirect fires continue support to the main defense elements until directed to disengage by the commander or leader and also exfiltrate.

Figure 7-12. Guerrilla defense of a complex battle position (example)

Command and Control of a CBP Defense

7-178. The irregular OPFOR leader will position himself where he can best command and control the defensive fight. C2 of a CBP is generally more difficult than that of an SBP because the defenders may be more dispersed. Insurgents or guerrillas operating in and from the CBP use secure communications such as couriers and wire.

Support of a CBP Defense

7-179. Support of a CBP can be provided from within the local insurgent or guerrilla organization and/or local resources in the relevant population. Some support may be allocated higher-level irregular OPFOR organizations or from a state or non-state sponsor. Specialized support such as SPF can be temporarily associated with insurgents or guerrillas in a CBP in order to provide training, materiel, and/or tactical and technical advice.

Reconnaissance

7-180. Reconnaissance assets observe avenues of approach key to providing early warning and allow the commander or leader time to defend or to exfiltrate personnel and resources from the CBP. Insurgents, guerrillas, and/or their active supporters embed themselves within local populations. The irregular OPFOR is less likely to engage in counterreconnaissance activities if these actions would reveal CBP location.

Armored Fighting Vehicles

7-181. Armored vehicles are not the norm in a CBP. When possessed by the irregular OPFOR, armored fighting vehicles and ad hoc fighting vehicles mounting heavy weapons are normally concealed and covered in hide positions. The commander or leader may retain armored vehicles as part of his reserve for quick response to contingencies in defense of his CBP.

Fire Support

7-182. CBPs are typically self-supporting in their defense. Fire support assets normally locate within the CBP but may also locate outside of the CBP perimeter to best employ specific fires. Defenders employ these fires to—
- Defeat enemy forces in the battle zone.
- Attrit enemy forces along avenues of approach near a CBP.
- Disrupt enemy use of landing zones in the vicinity of a CBP.
- Delay enemy forces to support withdrawal of the irregular OPFOR from the CBP.

Air Defense

7-183. Passive air defense is the norm for a CBP. Active air defense generally involves systems that do not emit an electromagnetic signature. Insurgents or guerrillas engage aerial targets on order of the commander or leader. An all-arms air defense concept involves using weapons of all personnel of a CBP to protect against fixed- and rotary-wing aircraft threats. However, guerrilla battalions typically have a limited MANPADS capability in their weapons company. When available, shoulder-fired MANPADS would likely be located in disruption or security element(s) that occupy fighting positions near a CBP oriented on probable enemy air avenues of approach. (See note under Air Defense in Support of an Ambush regarding additional air defense assets that may be available.)

Engineer-Like Capabilities

7-184. The irregular OPFOR conceals fighting and survivability positions using C3D techniques with locally available resources. Military-manufactured antipersonnel and/or antitank mines and/or IEDs are emplaced on or near likely enemy avenues of approach. Particular mines and/or IEDs are emplaced and secured with a cell or unit designated to arm and/or detonate the munition on order of the commander or leader.

7-185. The irregular OPFOR reinforces urban and/or rural terrain to channel enemy forces into kill zones on approaches to a CBP. Obstacles are generally more protective in nature than obstacles used near an SBP. Insurgents or guerrillas may have specialized talents for engineer-like countermobility and mobility tasks. When supported overtly or covertly by a state sponsor, SPF or regular forces may be associated with an insurgent or guerrilla organization in order to provide training, materiel, and advice.

Logistics

7-186. Logistics operations of a CBP are generally self-sustaining and blend into the local commerce and daily operations of a relevant population in the vicinity of the CBP. Provisioning a CBP with regular resupply is facilitated by active supporters of the insurgent or guerrilla organization. The commander or leader typically does not coerce local citizens to provide logistics support but can use such extortion when critical commodities are required to sustain the CBP defense.

7-187. Supply caches and safe houses are distributed throughout the urban or rural area near the CBP. Other supply caches and safe houses are located within the CBP perimeter.

INFOWAR

7-188. Elements from the CBP may attempt to integrate within any local communities for the purpose of gathering information, collecting intelligence, and disseminating INFOWAR themes to the local relevant population. Active supporters of the irregular OPFOR assist the commander or leader in keeping a low profile. INFOWAR activities may focus on downplaying the existence or significance of the CBP. Generally, the CBP will not conduct easily detectable INFOWAR activities that would call the enemy's attention to it. (See appendix A for additional information on irregular OPFOR INFOWAR capabilities.)

7-189. If the presence of a CBP cannot be hidden, INFOWAR may attempt to convince enemy forces that the defenders are friendly to them. It may attempt to convince leaders of the governing authority that the insurgents or guerillas are willing to affiliate and/or associate with them in levels of reconciliation. Other INFOWAR techniques can portray the enemy force and governing authority as a corrupt regime to further isolate them from the relevant population. The insurgents or guerrillas can claim and demonstrate themselves to be the population's protector. In some cases, senior irregular OPFOR leaders may conduct INFOWAR from a CBP to convince followers in other locations that they are still alive and leading their organizations in the struggle against the enemy. INFOWAR can include support to provide basic social and civic services to the relevant population that is not being provided by the governing authority.

7-190. Other INFOWAR techniques can result in the gradual acceptance by the relevant population to the point that members of the insurgent or guerrilla organization become informal or legitimate representatives in civil governance. This recognition can lead to election in a state's formal voting processes and/or establishing self-proclaimed semi-autonomous enclaves within a state with which the irregular OPFOR is in conflict. In either case, the irregular OPFOR ensures a significant INFOWAR campaign to weaken enemy support and strengthen its own support from a relevant population.

IRREGULAR OPFOR IN HYBRID THREAT TACTICS

7-191. The above section of this chapter shows examples of functional tactics in which insurgents and/or guerrillas employ some of the same types of offensive and defensive actions used by smaller tactical units of the irregular OPFOR (see TC 7-100.2). However, insurgents and/or guerrillas may also participate in larger-scale functional tactics when acting as part of the Hybrid Threat (HT). (See TC 7-100 for examples of such involvement.) They often work closely with SPF units in such roles.

7-192. One would expect the irregular OPFOR to play a relatively minor role in HT tactics, perhaps used as an economy-of-force measure when regular OPFOR units are not available. However, as examples in TC 7-100 show, insurgents and guerrillas can sometimes do things that regular military units cannot. For example, they are well suited to dispersed actions as part of a disruption force in a disruption zone (in offense or defense). However, their small size and ability to blend in with the local population also make it easier for irregular OPFOR units or cells to infiltrate into enemy-held positions. There, they may serve as enabling forces or elements, paving the way for exploitation by larger, regular military units, or they may play key roles as part of an exploitation force or other action force or element that accomplishes the overall mission of the HT action.

Note. Insurgents, guerrillas, or even criminals can perform acts of terrorism that complement functional tactics actions separate from or in conjunction with a larger HT force. (See chapter 6 for more details on terrorism.)

Appendix A
Information Warfare

The irregular OPFOR (including insurgents, guerrillas, and criminal elements) is trained to use adaptive techniques to defeat a superior opponent. Naturally it uses information warfare (INFOWAR) to obtain asymmetrical effects. These tactics are constantly evolving thanks to the exponential growth of networking and information technology. Increased accessibility to the Internet, commoditization of information, and unprecedented global awareness provide the irregular OPFOR with the ability to focus its INFOWAR activities on one of the enemies greatest weaknesses—his dependence on information communication technology in every aspect of his operations.

GENERAL CHARACTERISTICS

A-1. The irregular OPFOR, along with the regular OPFOR, defines *information warfare* as specifically planned and integrated actions taken to achieve an information advantage at critical points and times. Conducted in conjunction with offensive or defensive actions, INFOWAR is designed to exploit the weaknesses of the opponent's information systems.

GOALS

A-2. The primary goals of INFOWAR are to—
- Influence the enemy's decisionmaking through his collected and available information, information systems, and information-based processes.
- Retain the ability to employ friendly information and information-based processes and systems.

TACTICAL TASKS

A-3. The effects of INFOWAR can be multidimensional and at times hard to pinpoint. However, the irregular OPFOR highlights the following tasks and associated effects as critical to the application of INFOWAR at the tactical level:
- **Destroy.** Destruction tasks physically render the enemy's information systems ineffective. Destruction is most effective when timed to occur before the enemy executes a command and control (C2) function or when focused on a resource-intensive target that is hard to reconstitute. Neutralizing or destroying the opponent's information capability can be brought about by physical destruction of critical communications nodes and links.
- **Degrade.** Degradation attempts to reduce the effectiveness of the enemy's information infrastructure, information systems, and information collection means.
- **Disrupt.** Disruption activities focus on disrupting enemy observation and sensor capabilities at critical times and locations. Disruption impedes the enemy's ability to observe and collect information and obtain or maintain information dominance.
- **Deny.** Denial activities attempt to limit the enemy's ability to collect or disseminate information on the irregular OPFOR or deny his collection efforts.
- **Deceive.** Deception activities strive to mislead the enemy's decisionmakers and manipulate his overall understanding of OPFOR activities. Deception manipulates perception and causes disorientation among decisionmakers within their decision cycle.
- **Exploit.** Exploitation activities attempt to use the enemy's C2 or reconnaissance intelligence, surveillance, and target acquisition (RISTA) capabilities to the advantage of the irregular OPFOR. The irregular OPFOR also uses its various INFOWAR capabilities to exploit any enemy vulnerability.

Appendix A

- **Influence.** Influencing information affects an enemy's beliefs, motives, perspectives, and reasoning capabilities, in order to support irregular OPFOR objectives. This may be done through misinformation or by manipulating or "spinning" information.

INFOWAR ELEMENTS

A-4. Across the spectrum of competition, conflict, and war, the following elements are integrated when developing and implementing INFOWAR:

- **Electronic warfare (EW).** Activity conducted to control or deny the enemy's use of the electromagnetic spectrum, while ensuring its use by the irregular OPFOR.
- **Deception.** Measures designed to mislead the enemy by manipulation, distortion, or falsification of evidence to induce the enemy to react in a manner prejudicial to his interests.
- **Physical destruction.** Use of all types of weapons and explosives to destroy critical components of the enemy force, focusing on C2 nodes and RISTA sensors.
- **Protection and security measures.** A wide range of activities (including the elements of deception and EW) to enhance survivability and preserve combat power.
- **Perception management.** Measures aimed at creating a perception of truth that best suits irregular OPFOR objectives.
- **Information attack (IA).** Intentional disruption or distortion of *information* in a manner that supports accomplishment of the irregular OPFOR mission.
- **Computer warfare.** Attacks focused specifically on computer *systems, networks, and nodes.*

A-5. The seven elements of INFOWAR do not exist in isolation from one another and are not mutually exclusive. The overlapping of functions, means, and targets requires that they all be integrated into a single, integrated INFOWAR plan. However, effective execution of INFOWAR does not necessarily involve the use of all elements concurrently. In some cases, one element may be all that is required to successfully execute a tactical INFOWAR action. Nevertheless, using one element or subelement, such as camouflage, does not by itself necessarily constitute an application of INFOWAR.

A-6. The use of each element or a combination of elements is determined by the tactical situation and support to the overall objective. The size and sophistication of an enemy force also determines the extent to which the irregular OPFOR employs the various elements of INFOWAR. The commander or leader has the freedom to mix and match elements to best suit his tactical needs, within the bounds of guidance from higher authority. The resources and capabilities of the irregular OPFOR, and the appropriateness of the INFOWAR medium to the target, determine the choices and the extent to which the various INFOWAR elements are employed and integrated.

TOOLS

A-7. Tools for waging INFOWAR can include, but are not limited to—

- Conventional physical and electronic destruction means.
- Malicious software.
- Denial-of-service attacks.
- The Internet.
- The media.
- International public opinion.
- Communication networks.
- Various types of reconnaissance, espionage, and eavesdropping technologies.

The irregular OPFOR can employ INFOWAR tools from both civilian and military sources and from assets of third-party actors.

A-8. In some cases, the irregular OPFOR will conduct INFOWAR activities that are criminal in nature. For example, it may—

- Use known computer exploitation tactics such as identity theft and industrial espionage.
- Sell its computer warfare capabilities to create additional funding for its operations.

TARGETS

A-9. The OPFOR sees the targets of INFOWAR as an opponent's—
- Decisionmakers.
- Weapons and hardware.
- Critical information infrastructure.
- C2 system.
- Information and telecommunications systems.
- C2 centers and nodes.

Information links, such as transmitters, communication devices, and protocols, will be targeted. The irregular OPFOR is extremely adaptive and will employ the best option available to degrade, manipulate, influence, use, or destroy an information link. See table A-1 for typical examples of INFOWAR objectives and targets.

Table A-1. INFOWAR elements, objectives and targets

Elements	Objectives	Targets
Electronic Warfare (EW)	Exploit, disrupt, deny, and degrade the enemy's use of the electromagnetic spectrum.	C2 and RISTA assets and networks.
Deception	Mislead enemy decisionmakers. Cause confusion and delays in the decisionmaking process. Persuade the local population and/or international community to support irregular OPFOR objectives.	Key military decisionmakers. General population and international media sources and Internet sites.
Physical Destruction	Destroy the enemy's information infrastructures.	C2 nodes and links, RISTA assets, telecommunications, and power sources.
Protection and Security Measures	Protect critical assets.	Enemy RISTA assets.
Perception Management	Distort reality or manipulate information to support irregular OPFOR goals.	Enemy RISTA assets. Local populace and leaders. Media sources (international and domestic).
Information Attack (IA)	Alter or deny key information.	Decisionmakers and other users of information. Systems reliant on accurate information.
Computer Warfare	Disrupt, deny, or degrade the enemy's computer networks and information flow.	C2 and RISTA assets and networks.

WINDOWS OF OPPORTUNITY

A-10. To conduct successful action against a more powerful enemy force likely enjoying a technological overmatch, the irregular OPFOR must exploit windows of opportunity. Sometimes these windows occur naturally, as a result of favorable conditions in the operational environment (OE). Most often, however, the

Appendix A

irregular OPFOR will have to create its own opportunities. INFOWAR can help create the necessary windows of opportunity for any type of offensive or defensive action, by executing deception techniques, EW, perception management, and/or physical destruction.

A-11. Specific INFOWAR activities that can limit enemy force effectiveness can include—
- Deceiving enemy forces with recurring social and political activities by active supporters of the irregular OPFOR that appear non-threatening in the vicinity of the targeted objective.
- Exploiting enemy information collection with false intelligence provided by active supporters of the irregular OPFOR and/or insider threats.
- Disrupting enemy information collection through coercion and extortion of the relevant population to not cooperate with enemy forces.
- Influencing enemy forces with misinformation and/or manipulated claims against the governing authority to which the enemy forces belong.
- Demonstrating the inability of enemy forces to protect civilians, effectively defend civil and/or military facilities, and/or safeguard key representatives of the governing authority.
- Psychologically isolating the enemy force and governing authority from each other as well as from a relevant civilian population.
- Creating a positive impression of irregular OPFOR actions and objectives among local, regional, and/or global audiences.
- Influencing a local, regional, and/or global audience with near real-time media coverage of successful irregular OPFOR actions against enemy forces and the governing authority.
- Using media releases to indicate the increasing discontent in a relevant population due to the governing authority's actions and reactions affecting daily civil commerce and lifestyles.
- Using progressive INFOWAR activities to degrade the resolve of the enemy force and governing authority to continue counterinsurgent, counter-guerrilla, or anti-crime operations.

INFOWAR ORGANIZATIONS

A-12. Various types on non-state actors might be part of the irregular OPFOR, affiliated with it, or support it in some manner. Even those organizations that do not belong to the irregular OPFOR or support it directly or willingly could be exploited or manipulated to support its objectives. Irregular INFOWAR organizations can be semiformal in the case of insurgent organization or ad hoc in informal organizations such as guerrillas or criminal gangs. In most case, the irregular forces are organized to perform a specific mission. Their number and size will increase or decrease based on the tactical situation.

A-13. INFOWAR organizations conduct actions either in support of or independent of established irregular OPFOR units. The determining factors are base on those means that achieve their goals expeditiously.

Note. Special-purpose forces (SPF) can organize, train, and support parts of the irregular OPFOR (insurgents or guerrillas and possibly even criminal organizations) and conduct operations in conjunction with them. SPF missions can also include the use of INFOWAR.

A-14. Criminal elements will engage in INFOWAR activities to protect their operations. However, some criminal organization specialize in INFOWAR and use technology to illegally produce profits either by acquiring sensitive financial information or by selling hacking services and malware toolkits. (See chapter 4 for more information.) INFOWAR organizations will provide a variety of products and technical assistance to aligned irregular forces through direct action cells or through specialized INFOWAR cells.

INFOWAR Cells

A-15. In a local or higher insurgent organization, an INFOWAR cell plans, coordinates, and implements the organization's INFOWAR plan and provides guidance and assistance to the INFOWAR direct action cells whenever required. The INFOWAR cell may use a variety of lethal and nonlethal methods to influence the enemy and the population. Examples of such activities include but are not limited to—
- Propaganda such as night letters and disinformation campaigns.

- Incentivizing of destabilization activities such as criminal activities.
- Assassinations and sabotage.

IED and other terrorist attacks are often used to support the overall INFOWAR plan. These attacks may also be used in a manner to deceive the enemy and confuse the source of the attacks. The ultimate goal of the INFOWAR cell is to create and maintain the perception that the enemy cannot provide security and stability to the population.

A-16. Depending on the size, nature, and focus of the insurgent organization, the INFOWAR cell may be capable of several functions. Multiple functions and activities are necessary if the enemy has mounted his own INFOWAR campaign. Some example functions performed by the INFOWAR cell are—

- Information management (internal methods, links, and security).
- Media manipulation, psychological warfare (PSYWAR), and public affairs designed to influence the population.
- Communications (cyber embeds via Internet sites, propaganda videos, broadcast successes of direct action teams, or printing).
- Rumor control (misinformation and disinformation).
- Sabotage actions.
- Civic actions.
- Indoctrination training.
- Fund-raising (including international).
- Recruiting.
- Assistance in cyber-mining for intelligence.

All these functions are integrated to further short- and long-range goals.

INFOWAR DIRECT ACTION CELLS

A-17. An INFOWAR direct action cell supports the insurgent organization's INFOWAR plan and may or may not receive guidance from the INFOWAR cell. Direct action as the title indicates is the nature of its objective, purpose, and role. Improvised explosive devices (IEDs) and other terror attacks are often used to support the overall INFOWAR plan. These attacks may also be used in a manner to shift blame for the attack to their enemy.

A-18. Depending on the size, nature, and focus of the insurgent organization, an INFOWAR direct action cell may be capable of several functions. Some example functions performed by this cell are—

- Selective sabotage actions.
- Information management
- Media manipulation (misinformation and disinformation—PSYWAR).
- Communications (cyber embeds via Internet sites, propaganda and indoctrination videos, broadcast successes of other direct action teams).
- Civic actions.
- Assistance in the cyber-mining for intelligence.

All of these functions are integrated to further short- and long-range goals.

A-19. Some of the functions may require specialized expertise. For example, the media manipulation function (PSYWAR) may require expertise and/or advice from a cleric; a political, a tribal, ethic, or cultural leader; or other experts. Portions of the INFOWAR direct action cell may be dispersed and are assigned as the mission dictates. When countering the enemy's INFOWAR campaign, specific knowledge of stabilizing factors and systems is required. For example, if the enemy engages in reconstruction efforts the direct action cell may require specific knowledge to sabotage the operation.

A-20. The mission, combat conditions, and many other variables determine the configuration and composition of each direct action cell. Direct action cells do not have a fixed structure. Cell composition is not fixed and varies from cell to cell, mission to mission, and OE to OE. The structure, personnel, equipment, and weapons mix, all depend on specific mission requirements. Personnel select weapons

Appendix A

appropriate to the mission. Other equipment is added as required, such as computers, computer rigged vehicles, specialized antennas, and communications. Several members of the direct action cell may be hired INFOWAR specialists. Some functions can also be performed by personnel outside of the cell. There may be as few as one INFOWAR direct action cell to over 20 INFOWAR direct action cells, depending on the mission and other factors. Portions of the INFOWAR direct action cell are probably dispersed.

SUPPORT TO INFOWAR Cells

A-21. Irregular OPFOR organizations at times will compete for limited resources either from their higher headquarters or from an external supporting state or non-state actors. This competition appears to the outside observer as disjointed or lacking the discipline needed for unity of effort. However, individual leaders of the irregular OPFOR are allowed to develop lines of operation as it sees fit given the unique set of circumstances of its area of responsibility and the means at its disposal. When a particular tactic is proven to be effective, it will be replicated as necessary in order to exploit success, increase the perception of legitimacy for the irregular OPFOR cause, and to give the impression of progress. This sort of "groping in the dark" for a successful strategy means that the irregular OPFOR is able to experiment to find what works and to capitalize on effective tactics. The key is to open as many inroads as possible and to increase the likelihood for windows of opportunity for the irregular OPFOR to exploit the political, economic or social situation.

A-22. The operational variables will determine whether or not a local area will require all of the elements of tradecraft for a complete perception management campaign. Important issues such as regional conflicts, underprivileged and underrepresented populations, and the location of political, commercial or economic power centers, all have the potential to be targets of an INFOWAR campaign plan. Support may come in many forms from the general population. Online chat rooms, radio and television opinion shows and politically motivated groups may all provide support to the INFOWAR mission. As the irregular OPFOR assess the local environment and the enemy's center of gravity is determined, INFOWAR planners will target select groups, organizations, and individuals for a variety of activities.

A-23. Some INFOWAR missions may require cultural insights, organizational intelligence, and specialized expertise for use in an INFOWAR campaign. For example, the media manipulation function (PSYWAR) may require expertise and/or advice from a cleric; a political, a tribal, ethnic, or cultural leader; or other experts. Other functions could be used to obtain specific technical expertise such as the configuration of important infrastructure, industrial, economic, and financial systems to develop targets for cyber attack. INFOWAR activities will leverage the knowledge of individuals employed by these sectors of the operational environment.

INFOWAR SUPPORT TO IRREGULAR OPFOR

A-24. INFOWAR activities support all of the categories of irregular OPFOR—insurgents, guerrillas, and criminal elements. INFOWAR Cells provide the ways and means to conduct all elements of INFOWAR activities. INFOWAR actions in the form of perception management or propaganda are designed to garner support of the population for the irregular OPFOR. By establishing seemingly innocuous activities such as web logs (blogs) and news outlets, INFOWAR cells can establish inroads for C2 functions, targeting and other support functions. By targeting an information system's weakest link, (the user), INFOWAR cells can obtain information that provides inroads for further attacks by the irregular OPFOR or to by components of the Hybrid Threat. If the current governing authority has control over information technology, INFOWAR cells can conduct their operations from the sanctuary of a sympathetic nation.

A-25. INFOWAR activities provide support to the irregular OPFOR covertly in both times of peace and times of war. OPFOR INFOWAR organizations can provide a variety of products and technical assistance to aligned irregular organizations. INFOWAR cells also provide information technology resources such as servers and networks to the irregular OPFOR in order to protect INFOWAR activities and disassociate it from INFOWAR operations.

A-26. Some examples of INFOWAR support—
- Locations of media outlets and personalities both friendly and enemy.

- Accepted customs and norms and practices used to foster support for the irregular OPFOR.
- Inside access to government, commercial, and financial leaders and institutions that support the enemy.
- Locations of civilian and government communications infrastructure.
- Distracting of enemy forces through instantaneous civil disobedience such as flash mobs.
- Economic data vital to projecting required resources necessary to establish local rapport.
- Established networks for recruiting and mobilization of combatants and noncombatants.
- Enemy forces mission command tactics techniques and procedures, and command and control structure.

A-27. INFOWAR cells provide advisors and technological resources to the irregular OPFOR in either a covert or clandestine manner depending on the objective. Examples of such support could include either all or some of the elements of INFOWAR and range from hosting servers and managing networks for an insurgency or criminal element to supplying software and/or EW equipment to guerrillas for INFOWAR operations.

A-28. INFOWAR cells may or may not have the necessary resources to conduct research and development of INFOWAR technologies and to develop mature INFOWAR training and doctrine. In some cases, commercial-off-the-shelf technology satisfies the requirement for the irregular OPFOR. The INFOWAR cell may provide funds or equipment such as computers and network equipment, low-power GPS jammers, or software for use by other insurgents or by affiliated guerrillas or criminals. The benefit of using commercial information technology is that it is dual purpose and has less chance of raising suspicion when transporting it.

Perception Management Activities

A-29. Perception management involves measures aimed at creating a perception of truth or stability that best suits the irregular OPFOR objectives. Perception management integrates a number of widely differing activities that use a combination of true, false, misleading, or manipulated information. Targeted audiences range from enemy forces, to the local populace, to world popular opinion. At the tactical level, the irregular OPFOR seeks to undermine an enemy's ability to conduct combat operations through PSYWAR and other perception management activities aimed at deterring, inhibiting, and demoralizing the enemy and influencing civilian populations.

A-30. The various perception management activities include efforts conducted as part of—
- PSYWAR.
- Direct action.
- Public affairs.
- Media manipulation and censorship.
- Regional or international recruitment and/or fundraising for affiliated irregular forces.

A-31. Information communications technology and global dissemination of the 24-hour news cycle has empowered the irregular OPFOR to implement complex perception management activities public affairs to affect change, garner global support, and generally shape the OE to its purposes. INFOWAR activities engage in public affairs to involve interest groups and media outlets to gain influence and to obtain their objectives. The irregular OPFOR can enable political and civic leaders at all levels to engage the population to accept its ideology and support the irregular OPFOR.

A-32. Although the irregular OPFOR maintains that perception management activities conducted at the tactical level must be consistent with, and contribute to, the goals of the highest levels of organization, subordinates are allowed much discretion on the ways and means of achieving their perception management objectives. For example, forming a partnership with a charitable organization or a local business leader in order to obtain secure lines of communications, and a recruiting pool would be a natural extension of the strategy necessary to influence the local populace. If there is a religious or other ideological approach available, the irregular OPFOR will leverage this to establish drivers of social mobility such as educational organizations or business alliances. These organizations also provide the

Appendix A

irregular OPFOR a conduit for recruiting, indoctrination, and long-term influence within the target group. Additionally, the irregular OPFOR has the freedom to provide immediate assistance and disperse funds without delay during times of crisis or whenever there is an opportunity to meet a particular objective. This gives the irregular OPFOR the opportunity to be the so called "first with the most" in the struggle for hearts and minds.

Psychological Warfare

A-33. PSYWAR is a major contributor to perception management. Targeting the forces of the enemy, PSYWAR attempts to influence the attitudes, emotions, motivations, aggressiveness, tenacity, and reasoning of enemy personnel. In addition to the enemy's military forces, PSYWAR specialists also concentrate on manipulating the local population and international media in favor of the irregular OPFOR, turning opinion against the enemy's objectives. Planners focus special emphasis on highlighting enemy casualties and lack of success. They also highlight enemy mistakes, especially those that cause civilian casualties or damage civil infrastructure.

A-34. The irregular OPFOR skillfully employs media and other neutral players, such as nongovernmental organizations (NGOs), to further influence public and private perceptions. However, if the irregular OPFOR perceives the presence of NGOs to be detrimental to its objectives, it can be extremely effective in hindering their efforts to provide humanitarian assistance, thus discrediting them.

Public Affairs

A-35. Local partnerships and projects are regarded as enhancing the strategic and operational goals of the irregular OPFOR but are not necessarily prescribed by the higher command. The objective is to provide a working solution that is culturally acceptable to the target population and does not compromise the core the irregular OPFOR. The irregular OPFOR seeks to integrate its activities into the target society and does this by providing the essential services for everyday life. Through cultural acceptance and shared goals, INFOWAR operators are able to develop trust and loyalty among the society and create opportunity for future projects and the perception of social mobility. Other examples of grassroots assistance given to a disenfranchised segment of the population by the irregular OPFOR could include—

- Establishment or purchase of a local business or industry in order to buy influence, generate funds for irregular OPFOR activities, and provide access to lines of communications.
- Cash payments to victims of both natural and manmade disasters.
- Support to religious, educational, or charitable institutions for public relations purposes, and recruitment.
- Provision of public services such as welfare, disaster response, or law and order services in order to delegitimize the existing government.
- Monetary support to religious, political, academic, or business leaders who are willing to support the irregular OPFOR cause.
- Establishing a parallel legal process where the population can obtain a just resolution for disputes without unwanted corruption by external values.

A-36. If properly employed, the results of perception management activities become ingrained into everyday life and can be viewed as a positive force to the targeted population. The targeted population gets the services denied to them by the current governmental structure, while the irregular OPFOR is able to move freely among the population and establish a support structure for future operations. Perception management activities are regarded by the enemy as propaganda, despite the fact that the irregular OPFOR enjoys more influence over the population than the existing government does. The irregular OPFOR is able to maintain contact with the target population in an overt way that further legitimizes its presence. By providing opportunities for education, work and charity, the irregular OPFOR receives in return loyalty and support for its cause. The irregular OPFOR may adopt a long-term strategy that allows it to fully integrate into all aspects of society. The fact that it administers resources and services that are unavailable to the targeted population increases its influence and makes affiliation with its cause a desirable goal for the target population.

A-37. Providing footage of perceived abuses of power by the enemy to the public helps INFOWAR activities make a case for allowing the irregular OPFOR to establish an ideological foothold within the state and provides the moral support needed to increase the resistance to the government.

Disaster Response

A-38. Response to disaster, whether natural or manmade, is viewed by the irregular OPFOR as another opportunity to gain influence and support in a region. Human suffering on a large scale sets the conditions for chaos and an overextension of the state's resources. In many regions, disaster-relief services may be inadequate, and there is usually an inordinate amount of suffering before any assistance becomes available. Because of its access to resources and support systems that are outside the government's bureaucratic structure, the irregular OPFOR can enable a more comprehensive response to natural disasters in certain targeted areas. In some cases, it will augment the current regime's disaster-relief efforts and attempt to integrate and legitimize its role in assisting the population. In other situations, it will supplant the existing structure and outperform the competition. The goal is to be the first with the most in terms of aid and assistance. Disaster response efforts may include—
- Evacuation of personnel from threatened areas.
- Provision of humanitarian relief such as food and temporary shelter.
- Long-term plans to rebuild structures destroyed by the disaster.
- Cash payments to victims to pay for immediate needs or to compensate a loss.

A-39. The irregular OPFOR can couple these services—including grassroots activism, social services and disaster response—with a political message and an intimidating presence. These combined methods allow the irregular OPFOR to establish its legitimacy, separate its enemies from its friends, and build support among the population while establishing inroads for future operations.

Media Activities

A-40. Irregular OPFOR will use all available means to exploit the media. These include both traditional forms such as television and newspaper outlets and as "new" media such as social networks and other computerized outlets. Benefits to the irregular OPFOR include the ability to connect with sympathizers for recruiting and resourcing, channeling responses through emotive themes, and promulgating a narrative of events that is more conducive to its objective.

A-41. The barriers to obtaining and using the media have diminished significantly with the ubiquitous expansion of the Internet. Media resources are no longer restricted to a limited number of corporations or individuals. Individuals with the technical knowledge can leverage a variety of commercially available technologies and reach a large audience. In some cases video releases or newscasts produced exclusively within the new media are used by traditional media outlets and are given more attention than the official account of events. Unburdened by a multi-step approval process, irregular OPFOR can manipulate the media to their political and military advantage. Oftentimes it is not the most accurate or properly vetted message that attracts attention, but rather the first to hit the media.

A-42. Having the ability to appeal or relate to the broadest possible audience within the cultural context of the OE, means that the irregular OPFOR has an intimate understanding of the target audience including social and cultural factors such as race issues, economic status and education levels.

A-43. The irregular OPFOR uses its knowledge of technology and culture to craft a message that is both targeted and suggestive. It also uses media operations as a front for reconnaissance and intelligence operations. Some of the techniques available to the irregular OPFOR are —
- Producing pamphlets, signs, and banners to spread their message usually through the use of images for illustration.
- Using radio television and other media in a fashion that exploits freedom of speech laws in order to operate on the fringe of social norms without being illegal.

Computer Warfare Information Attack

A-44. OPFOR information warfare (INFOWAR) units engage in a combination of computer warfare, information attack and perception management to establish and maintain information dominance in cyberspace by using both direct and indirect attacks on their adversary's computer networks, information systems and online media. Computer warfare and information attack targets both the networks and the information enemy elements of national power need to function daily basis and conduct warfare. INFOWAR brings to bear the information oriented elements of national power to decisively defeat the enemy.

A-45. The complex nature of the information environment provides the conditions for both direct and indirect INFOWAR attacks and presents an opportunity for INFOWAR tactics and techniques to augment traditional military operations at all levels of command from tactical to the strategic. Modern militaries rely on information systems and automation as well as worldwide connectivity to project power on a global scale. These information systems are used in command and control; logistics support; surveillance and reconnaissance; and for global positioning and navigation satellites which provide both navigation and precision targeting data.

A-46. The direct approach to INFOWAR is a concerted operation designed to attack a specific target whether it is key infrastructure, a civil institution's databases or critical command and control nodes. While the indirect approach could involve the constant probing of the adversary's weaknesses and multiple limited engagements designed to wear down electronic defenses, win decisively in the propaganda campaign, or to identify vulnerabilities for future operations.

Providing Computer Warfare Tools and Services

A-47. Hosting services are the key to supporting INFOWAR activities in cyberspace. The INFOWAR unit provides access for a variety of services and is a clearing house for exchanging successful techniques and data on potential targets in chat rooms and on blogs. The irregular OPFOR can use these services to email or embed links to servers outside the area of operations that can be used for recruiting, exchanging intelligence and propaganda, and launching attacks.

A-48. Propaganda is still a key component of the INFOWAR campaign. Redirecting Internet users to OPFOR themed websites will have the effect of not only spreading the irregular OPFOR's message but giving the enemy and his supporters the impression that the irregular OPFOR has compromised every aspect of his society.

Developing Paralell Networks

A-49. Providing to the irregular OPFOR a C2 network is an effective way for the INFOWAR activity to maintain situational awareness and gather intelligence on the state's infrastructure and elements of power. A dedicated communications network can also be used to recruit and mobilize large groups for specific actions such as rallies, protests, riots, or attacks.

A-50. The irregular OPFOR posing as commercial or humanitarian venture, such as an NGO, a legitimate business, or another noncombatant entity, can establish network connectivity in a contested area of interest. This can be accomplished at another level by providing telecommunications infrastructure and switching equipment to the state's telecommunications company with embedded backdoor access to the input-output systems of the equipment.

A-51. This will give the INFOWAR activity free access to important elements of the nation's telecommunications network. In other cases, after hostilities have begun, telecommunications equipment can be replaced or reconfigured after a key node is physically seized and repurposed for manipulation and use by the irregular OPFOR.

A-52. The irregular OPFOR has the ability to seize enemy civil telecommunications infrastructure and use it for their purposes. Existing switching equipment can be integrated with that of their own and used with existing lines and wireless towers to extend the range and services for the irregular OPFOR C2 network. Once implemented the irregular OPFOR can commercialize its services. As more subscribers are added, the network becomes a source of intelligence by logging calls made and received by persons of interest. It can also be a source of revenue to fund future actions.

SPF and INFOWAR

A-53. INFOWAR capabilities provide clear advantages for SPF missions, due to the fact that SPF will always conduct missions in small teams or detachments and will use INFOWAR capabilities and techniques to maximize the effectiveness of their small numbers. INFOWAR capabilities have the potential to increase situational awareness, facilitate deception, and disrupt the enemy's decision making process.

A-54. SPF conducts an array of operations in support of the INFOWAR plan. Examples include—
- Perception Management.
- Diversionary Measures.
- Sniper.
- Indirect Fire.
- Air Defense (denial of airspace and enemy claims of air supremacy and/or superiority).
- Electronic Warfare.

Note. SPF can recruit, organize, train, advise, and support local insurgents or guerrillas and possibly even criminal organizations and conduct (or lead) operations in conjunction with them. SPF personnel may fight alongside such affiliates or assist them to prepare for offensive actions, diversionary measures, INFOWAR activities or other missions. In some cases, the SPF will not only advise and assist but actually control (command) the irregular forces as a surrogate force. SPF missions (and those of their affiliates or surrogates) can include the use of terror tactics. See TC 7-100.2 Opposing Force Tactics Chapter 7 INFOWAR, and Chapter 15 Special Purpose Forces for more information.

A-55. The nature of the shared goal or interest determines the tenure and type of relationship and the degree of affiliation. For example, the affiliation of an SPF detachment with criminal (or guerrilla) organizations is dependent only on the needs of the criminal (or guerrilla) organization or on the needs of the SPF at a particular time. The relational dynamics of SPF units are very fluid and apt to change from one day to the next. Shifts in affiliations may in turn cause adjustments in the SPF task organization to accommodate these changes.

Direct Action Team

A-56. A Direct Action Team is embedded in most INFOWAR units to provide support to the PSYWAR company. The teams specialize in infantry antiarmor style attacks; antiarmor ambushes; hit-and-run attacks against armored and/or hardened or rear area targets. Usually one Direct Action Teams, each broken into three attack elements of 4 persons each. The teams will probably be supported or augmented by different types of SPF specialty teams such as Sniper Teams or Sapper Teams. The SPF Teams (UAV) may be used to acquire reconnaissance information on targets or facilities.

A-57. SPF team members are cross-trained in the use of all equipment, weapons, and vehicles assigned to the company. The equipment and weapons mix is determined by the mission. SPF Soldiers speak several languages and are able to interact with the local populace for perception management and or PSYWAR missions. All SPF teams may be augmented by other SPF personnel, weapons, and/or equipment. The team serves as a standard SPF Team when not required for INFOWAR.

A-58. The SPF team leader coordinates all combat activities as well as reconnaissance and jamming activities, operations, employment, monitoring, direction finding, collection, and reporting. This includes GSR, sensor sets, remote sensor monitoring, and observation.

Signal Team

A-59. A single small SPF signal team can provide long-range communications support for guerrilla units up to battalion size. A full SPF signal team can do the same for a brigade-size unit. Teams can also support insurgent operations. This team may also serve in a signals reconnaissance collection role. In the collection role, the signal equipment is exchanged one-for-one with communications intercept and direction finding equipment. Each team then becomes a communications intercept and direction finding unit.

Sniper Team

A-60. The snipers (7.62-mm or .50 cal) in each attack element provide covering fires. The 35-mm AGL-L and the 7.62-mm GP MG engage personnel when they exit the armored vehicles, pin down the supporting infantry allowing the ATGM/ATGL gunner to engage the armored vehicle, and provide covering fires for the ATGM and ATGL gunners. Sniper operations are useful to create hesitancy among enemy forces and to disrupt the enemy decision making process by forcing enemy commanders to account for sniper counter sniper operations.

Glossary

SECTION I – ACRONYMS AND ABBREVIATIONS

AKO	Army Knowledge Online
AOR	area of responsibility
AP	antipersonal
AR	Army regulation
AT	antitank
ATGL	antitank grenade launcher
ATGM	anitank guided missile
AUTL	Army Universal Task List
BP	battle position
C2	command and control
C3D	camouflage, cover, concealment, and deception
CBP	complex battle position
CBRN	chemical, biological, radiological, nuclear
CI	counterintelligence
CP	comand post
CR	counterreconnaissance
CS	combat support
CSS	combat service support
DA	direct action
DODD	Department of Defense Directive
EFP	explosive formed projectile
EW	electronic warfare
FM	field manual
G	guerrilla [in guerrilla unit symbol]
GANG	criminal (generic) [in crimainal organization symbol]
GPS	global positioning system
HEAT	high explosive antitank
HK	hunter-killer
HRO	humantarian relif organization
HT	Hybrid Threat
HUMINT	human intelligence
HVT	high-value target
I	insurgent [in insurgent organization symbol]
IA	information attack
IED	improvised explosive device
INFOWAR	information warfare
INS	internal security
ISF	internal security forces
JP	joint publication
km	kilometer(s)
LNG	liquefied natural gas

Glossary

LOC	line of communications
LVCG	live, virtual, constructive, and gaming
LZ	landing zone
MANPADS	man-portable air defense system
m	meter(s)
mm	millimeter(s)
MRL	multiple rocket launcher
NGO	nongovernmental organization
OE	operational environment
OP	observation post
OPFOR	opposing force (AR 350-2)
OPSEC	operational security
PSC	private security contractor
PMESII-PT	political, military, economic, social, information, infrastructure, physical environment, and time
PSYWAR	psychologcial warfare
POL	petroleum, oils, and lubricants
RISTA	reconnaissance, intelligence, surveillance, and target acquisition
SAF	small arms fire
SAM	surface-to-air missile
SATCOM	satellitte communications
SBP	simple battle position
SIGINT	signals intelligence
SPF	special-purpose forces
SSM	surface-to-surface missile
SVIED	suicide vest improvised explosive device
SVBIED	suicide vehicle borne improvised explosive device
TC	training circular
TCP	traffic control post
TIC	toxic industrial chemical
TIM	toxic idustrial material
TRADOC	Training and Doctrine Command [US Army]
TRISA	TRADOC G-2 Intelligence Support Activity [US Army]
TTP	tactics, techniques, and procedures
US	United States
UAV	unmanned aerial vehicle
UJTL	Universal Joint Task List
UN	United Nations
VBIED	vehicle borne improvised explosive device
WMD	weapon of mass destruction

Glossary

SECTION II – TERMS

adherent
> An individual or one or more unit, organization, or cell that forms collaborative relationships with, acts on behalf of, or is otherwise *inspired*, without any requirement of allegiance to another organization, to take action in support of the goals and objectives of another irregular force [such as al-Qa'ida—the organization and the ideology—including support and/or acts of violence]. (As adapted for TC 7-100.3 from *U.S. National Strategy for Counterterrorism*)

affilitate
> An individual or one or more unit, organization, or cell that *aligns with* another individual, unit, organization, or cell to influence, support, and/or act in concert with for mutual benefit. No command relationship exists necessarily between an affiliate and the unit, organization, or cell in whose area of responsibility it operates; however, localized operations of the affiliate may support the area, regional, or global agenda of an organization [such as al-Qa'ida]. Affiliates are typically nonmilitary or paramilitary individuals or groups. In some cases, affiliated forces may receive support from regular military forces as part of an agreement under which they cooperate. (As adapted for TC 7-100.3 from *U.S. National Strategy for Counterterrorism*)

antiterrorism
> Defensive measures used to reduce the vulnerability of individuals and property to terrorist acts, to include limited response and containment by local military and civilian forces. Also called AT. (JP 3-07.2)

ambush
> A surprise attack from a concealed position used against moving or temporarily halted targets in order to destroy or capture personnel and/or supplies; harrass and demoralize the enemy; delay or block movement of personnel and supplies; and/or canalize enemy movements. (TC 7-100.2)

area defense
> A defensive operation designed to achieve tactcial decision by forcing the enemy's offensive operations to culminate before the enemy can achieve his objective, and/or by denying the enemy his objectives while preserving combat power until decision can be achieved through strategic operations and operational mission accomplishment. (TC 7-100.2)

area of reposnibility
> The geographical area and associated airspace, as defined by the Opposing Force (OPFOR), within which a commander has the authority to plan and conduct combat operations. An AOR is bounded by a *limit of responsibility* (LOR) beyond which the organization may not operate or fire without coordination through the next-higher headquarters. AORs may be linear or nonlinear in nature. Linear AORs may contain subordinate nonlinear AORs and vice versa. (TC 7-100.2)

assault
> An attack that destroys an enemy force through firepower and the physical occupation and/or destruction of the enemy position. (TC 7-100.2)

attack
> An offensive operation that destroys or defeats enemy forces, seizes and secures terrain, or both destroys or defeats enemy forces and seizes and secures terrain. (TC 7-100.2)

battle
> A battle consists of a set of related engagements that last longer and involve larger forces than an engagement. (ADRP 3-90)

battle positon
> A defensive location oriented on a likely enemy avenue of approach. (TC 7-100.2)

Glossary

combating terrorism
　Actions, including antiterrorism (defensive measures taken to reduce vulnerability to terrorist acts) and counterterrorism (offensive measures taken to prevent, deter, and respond to terrorism), taken to oppose terrorism throughout the entire threat spectrum. Also called CbT. See antiterrorism and counterterrorism. (See also JP 3-26)

complex battle position
　A defensive location designed to employ a combination of complex terrain, C3D, and engineer effort to protect the unit(s) within it from detection and attack while denying their seizure and occupation by the enemy. (TC 7-100.2)

condition
　Those variables of an operational environment or situation in which a unit, system, or individual is expected to operate and may affect performance. See also joint mission-essential tasks. (JP 1-02)

counterterrroism
　Operations that include the offensive measures taken to prevent, deter, preempt, and respond to terrorism. Also called CT. (JP 3-26)

deception
　Those measures designed to mislead the enemy by manipulation, distortion, or falsification of evidence to induce him to react in a manner prejudicial to his interests. (JP 1-02; JP 3-13.4)

enemy
　A party identified as hostile against which the use of force is authorized. (ADRP 3-0) [The *enemy* in the context of TC 7-100.3 is any hostile party to the irregular opposing force (TC 7-100.3)]

enemy combatant
　In general, a person engaged in hostilities against the United States or its coalition partners during an armed conflict. The term "enemy combatant" includes both "lawful enemy combatants" and "unlawful enemy combatants." (U.S. Department of Defense Directive (DODD) 2310.01E, *The Department of Defense Detainee Program*, September 5, 2006)

engagement
　A tactical conflict, usually between opposing, lower echelon maneuver forces. (ADP 3-0)

guerrilla
　A typically indigenous individual within a irregular unit structure organized along military lines in order to conduct military and paramilitray operations in enemy-held, hostile, or denied territory. TC 7-100.3 typically identifies guerrillas within the context of guerrilla units. [As adapted to TC 7-100.3 from JP 1-02/No approved JP 1-02 definition]

guerrilla force
　A group of irregular, predominantly indigenous personnel organized along military lines to conduct military and paramilitary operations in enemy-held, hostile, or denied territory. (JP 1-02; JP 3-05)

hybrid threat
　The diverse and dynamic combination of regular forces, irregular forces, and/or criminal elements all unified to achieve mutually benefiting effects. (ADRP 3-0; see also TC 7-100)

improvised explosive device
　A weapon that is fabricated or emplaced in an unconventional manner incorporating destructive, lethal, noxious, pyrotechnic, or incendiary chemicals designed to kill, destroy, incapacitate, harass, deny mobility, or distract. Also called IED. (JP 1-02)

information operations
　The integrated employment of the core capabilities of electronic warfare, computer network operations, psychological operations, military deception, and operations security, in concert with specified supporting and related capabilities, to influence, disrupt, corrupt or usurp adversarial human and automated decision making while protecting the irregular opposing force human and automated decision making. Also called IO. [As adapted for TC 7-100.3 from JP 3-13]

Glossary

intergovernmental organization
 An organization created by a formal agreement between two or more governments on a global, regional, or functional basis to protect and promote national interests shared by member states. (ADRP 3-0)

irregular forces
 Armed individuals or groups who are not members of the regular armed forces, police, or other internal security forces. (JP 1-02; JP 3-24)

irregular warfare
 A violent struggle among state and non-state actors for legitimacy and influence over relevant population(s). Irregular warfare favors indirect and asymmetric approaches, though it may employ the full range of military and other capacities, in order to erode an adversary's power, influence, and will. (JP 1-02)

insurgent
 An individual organized within an irregular insurgent organization structure that uses subversion and/or violence in order to overthrow or force change of a governing authority. TC 7-100.3 typically identifies insurgents within the context of cells. [As adapted for TC 7-100.3 from JP 1-02 and JP 3-24/no approved JP 1-02 defintion.]

kill zone
 A designated area on the battlefield where the opposing force (OPFOR) plans to destroy an enemy using obstacles and massed fires of direct and indirect weapon systems. (TC 7-100.2) [As adapted for TC 7-100.3 from FM 3-21.8, ATTP 3-06.11, ATTP 3-21.9.]

lawful enemy combatant
 Lawful enemy combatants, who are entitled to protections under the Geneva Conventions, include members of the regular armed forces of a State party to the conflict; militia, volunteer corps, and organized resistance movements belonging to a State party to the conflict, which are under responsible command, wear a fixed distinctive sign recognizable at a distance, carry their arms openly, and abide by the laws of war; and members of regular armed forces who profess allegiance to a government or an authority not recognized by the detaining power. (Department of Defense Directive (DODD) 2310.01E, *The Department of Defense Detainee Program*, 5 September, 2006)

maneuver defense
 A defensive operation designed to achieve tactcial decision by skillfully using fires and manuever to destroy key enemy elements of the enemy's combat system and deny enemy forces their objective, while preserving the friendly force. (TC 7-100.2)

militia
 An organization which generally refers to citizens trained as soldiers (as opposed to professional soldiers), but applies more specifically to a state-sponsored militia that is part of the state's armed forces but subject to activation only in an emergency. To avoid confusion, the TC 7-100 series uses *militia* typically in the latter sense. Irregular forces might be referred to or declare itself as a "militia;" however, the term *militia* is not typically used to describe gurrilas, insurgents, or criminals associated with opposing forces. (TC 7-100.2; TC 7-100; TC 7-100.3)

mercenary
 An individual who acts individually or acts a member of a formed group and volunteers from recruitment locally or abroad in order to fight in an armed conflict; is operating directly in the hostilities; is motivated by the desire for private gain, are promised, by or on behalf of a party to the conflict, material compensation substantially in excess of that promised or paid to the combatants of similar rank and functions in the armed forces of that party; is neither a national of a party to the conflict nor residents of territory controlled by a party to the conflict; is not a member of the armed forces of a party of the conflict; and, is not on official military duty representing a country that is not involved in the conflict such as a legitimate loan service or training appointment between. (*Geneva Conventions IV*)

nongovernmental organization
 A private, self-governing, not-for-profit organization dedicated to alleviating human suffering; and/or promoting education, health care, economic development, environmental protection, human rights, and

Glossary

conflict resolution; and/or encouraging the establishment of democratic institutions and civil society. NGO. (JP 3-08)

objective
The clearly defined, decisive, and attainable goal toward which every operation is directed. A location on the ground used to orient operations, phase operations, facilitate changes of direction, and provide for unity of effort. (ADRP 1-02)

operational area security
A form of security operations conducted to protect friendly forces. Forces engaged in area security operations focus on the force, installation, route, area, or asset to be protected. (ADRP 3-37)

operational environment
A composite of the conditions, circumstances, and influences that affect the employment of capabilities and bear on the decisions of the commander. (JP 3-0)

operational variables
Those interrelated aspects of an operational environment, both military and nonmilitary, that differ from one operational environment to another and define the nature of a particular operational environment. The eight operational variables are political, military, economic, social, information, infrastructure, physical environment, and time (PMESII-PT) (ADRP 3-0)

opposing force
A plausible, flexible military and/or paramilitary force representing a composite of varying capabilities of actual worldwide forces, used in lieu of a specific threat force for training and developing US forces (AR 350-2).

paramilitary
An irregular individual belonging to forces or groups distinct from the regular armed forces of any country, but resembling them in organization, equipment, training, or mission. [As adapted for TC 7-100.3 from JP 1-02 and JP 3-24]

paramilitary forces
Forces or groups distinct from the regular armed forces of any country, but resembling them in organization, equipment, training, or mission. (JP 1-02; JP 3-24)

procedure
Standard, detailed steps that prescribe how to perform specific tasks. See also tactics; techniques. (JP 1-02)

propaganda
Any form of adversary communication, especially of a biased or misleading nature, designed to influence the opinions, emotions, attitudes, or behavior of any group in order to benefit the sponsor, either directly or indirectly. (JP 1-02)

protection
The preservation of the effectiveness of mission-related military and nonmilitary personnel, equipment, facilities, information, and infrastructure deployed or located within or outside the boundaries of a given operational area. (ADRP 3-37)

raid
An attack against a stationary target for the purpose of its capture or destruction that culminates in the withdrawal of the raiding dorce to safe territory. (TC 7-100.2)

reconnaissance attack
A tactical offensive action that locates moving, dispersed, or concealed enemy elements and either fixes or destroys them; and/or gain information about the enemy's location, dispositons, military capabilities, and/or intentions. (TC 7-100.2)

Glossary

red team
An organizational element comprised of trained and educated members that provide an independent capability to fully explore alternatives in plans and operations in the context of the operational environment and from the perspective of adversaries and others. (JP 1-02)

resistance movement
An organized effort by some portion of the civil population of a country to resist the legally established government or an occupying power and to disrupt civil order and stability. (JP 1-02; JP 3-05)

risk management
The process of identifying, assessing, and controlling risks arising from operational factors and making decisions that balance risk cost with mission benefits. (JP 1-02)

rules of engagement
Directives issued by competent military authority that delineate the circumstances and limitations under which [United States] forces will initiate and/or continue combat engagement with other forces encountered. Also called ROE (JP 1-04; ADRP 3-0)

simple battle position
A defensive location oriented on the most likely enemy avenue of approach. (TC 7-100.2)

standard
A satisfactory level of performance for a task and condition for each individual and collective task to ensure that the individual or organization meets mission requirements. When no standard exists, the commander or organizational leader establishes a standard and the next higher commander or leader approves it. All training conducted by the unit, organization, or cell is assessed against the commander's or leader's intent for the training event, mission, and/or published doctrinal standards. See also TC 7-101, Appendix A. [As adapted for TC 7-100.3 from ADP 7-0]

tactics
The employment and ordered arrangement of forces in relation to each other. See also procedures; techniques. (JP 1-02)

task-organizing
The act of designing an operating force, support staff, or sustainment package of specific size and composition to meet a unique task or mission. (ADRP 3-0)

techniques
Non-prescriptive ways or methods used to perform missions, functions, or tasks. See also procedures; tactics. (JP 1-02)

terrorist
An individual who commits unlawful acts of violence or threat of violence to instill fear and coerce governments and societies in pursuit of political, religious, or ideological objectives. TC 7-100.3 typically identifies terrorists in the context of cells. [As adapted to TC 7-100.3 from JP 1-02/no approved JP 1-02 defintion.]

terrorism
The use of unlawful violence or threat of unlawful violence to inculcate fear; intended to coerce or to intimidate governments or societies. Terrorism is often motivated by religious, political, or other ideological beliefs and committed to the pursuit of goals that are usually political. (JP 1-02 and JP 3-07.2)

threat
Any combination of actors, entities, or forces that have the capability and intent to harm the United States forces, United States national interests, or the homeland. (ADRP 3-0) *Note.* Irregular opposing forces can apply this definition to actors, entities, or forces that have the capability and intent to harm irregular opposing force goals and objectives.

threat analysis
In antiterrorism, a continual process of compiling and examining all available information concerning potential terrorist activities by terrorist groups which could target a facility. A threat analysis will

Glossary

review the factors of a terrorist group's existence, capability, intentions, history, and targeting, as well as the security environment within which friendly forces operate. Threat analysis is an essential step in identifying probability of terrorist attack and results in a threat assessment. See also antiterrorism. (JP 1-02) *Note*. Irregular opposing forces can apply this definition to actors, entities, or forces that have the capability and intent to harm irregular opposing force goals and objectives.

traditional warfare

A form of warfare between the regulated militaries of states, or alliances of states, in which the objective is to defeat an adversary's armed forces, destroy an adversary's war-making capacity, or seize or retain territory in order to force a change in an adversary's government or policies. (DOD Directive 3000.07, *Irregular Warfare (IW)*, December 1, 2008)

transnational threat

Any activity, individual, or group not tied to a particular country or region that operates across international boundaries and threatens [United States] national security interests. (JP 3-26) *Note*. Irregular opposing forces can apply this definition to actors, entities, or forces that have the capability and intent to harm irregular opposing force goals and objectives.

unconventional warfare

A broad spectrum of military and paramilitary operations, normally of long duration, predominantly conducted through, with, or by indigenous or surrogate forces who are organized, trained, equipped, supported, and directed in varying degrees by an external source, and includes but is not limited to, guerrilla warfare, subversion, sabotage, intelligence activities, and unconventional assisted recovery. (JP 1-02)

unlawful enemy combatant

Persons not entitled to combatant immunity, who engage in acts against the United States or its coalition partners in violation of the laws and customs of war during an armed conflict. For purposes of the war on terrorism, the term *unlawful enemy combatant* is defined to include, but is not limited to, an individual who is or was part of or supporting Taliban or al Qaeda forces or associated forces that are engaged in hostilities against the United States or its coalition partners. (Department of Defense Directive (DODD) 2310.01E, *The Department of Defense Detainee Program*, September 5, 2006)

weapon of mass destruction

Chemical, biological, radiological, or nuclear weapons capable of a high order of destruction or causing mass casualties and exclude the means of transporting or propelling the weapon where such means is a separable and divisible part from the weapon. Also called WMD. (JP 3-40) *Note*. WMD effects can be caused by other means such as high-yield or low-yield explosives.

References

DOCUMENTS NEEDED
These documents must be available to the intended users of this publication.
ADP 1-02. *Operational Terms and Graphics*. 24 September 2013.
ADP 7-0. *Training Units and Developing Leaders*, 23 August 2013.
ADRP 1-02. *Operational Terms and Graphics*. 24 September 2013.
AR 350-2. *Opposing Force Program*. 9 April 2004.
DOD MIL-STD-2525-C. *Common Warfighting Symbology*. 17 November 2008.
JP 1-02. *Department of Defense Dictionary of Military and Associated Terms*. Available online at http://www.dtic.mil/doctrine/jel/doddict/.
TC 7-100. *Hybrid Threat*. 26 November 2010.
TC 7-100.2. *Opposing Force Tactics*. 9 December 2011.
TC 7-101. *Exercise Design*. 26 November 2010.

READINGS RECOMMENDED
These sources contain relevant supplemental information.
ADRP 3-0. *Unified Land Operations*. 16 May 2012.
ADRP 3-05. *Special Operations*. 31 August 2012.
ADRP 3-07. *Stability*. 31 August 2012.
ADRP 3-37. *Protection*. 31 August 2012.
ADRP 3-90. *Offense and Defense*. 31 August 2012.
ADRP 6-0. *Mission Command*. 17 May 2012.
ADRP 6-22. *Army Leadership*. 1 August 2012.
AR 11-33. *Army Lessons Learned Program (ALLP)*. 17 October 2006.
AR 381-11. *Intelligence Support to Capability Development*. 26 January 2007.
ATTP 3-06.11. *Combined Arms Operations in Urban Terrain*. 10 June 2011.
ATTP 3-21.9. *SBCT Infantry Platoon and Squad*. 8 December 2010.
DODD 2310.01E. *The Department of Defense Detainee Program*. 5 September 2006.
DODD 3000.07. *Irregular Warfare*. 1 December 2008.
FM 3-21.8 *The Infantry Rifle Platoon and Squad*. 28 March 2007.
FM 3-22. *Army Support to Security Cooperation*. 22 January 2013.
FM 3-24, *Counterinsurgency*. 15 December 2006.
FM 3-24.2. *Tactics in Counterinsurgency*. 21 April 2009
FM 3-37.2. *Antiterrorism*. 18 February 2011.
FM 7-15. *The Army Universal Task List*. 27 February 2009.
FM 7-100.1, *Opposing Force Operations*. 27 December 2004.
FM 7-100.4, *Opposing Force Organization Guide*. 3 May 2007.
Geneva Conventions. http://www.loc.gov/rr/frd/Military_Law/Geneva_conventions-1949.html
JP 3-05. *Special Operations*. 18 April 2011
JP 3-07.2. *Antiterrorism*. 24 November 2010.
JP 3-08. *Interorganizational Coordination During Joint Operations*. 24 June 2011.
JP 3-13. *Information Operations*. 27 November 2012.
JP 3-13.4. *Military Deception*, 26 January 2012.

JP 3-24. *Counterinsurgency Operations.* 5 October 2009.
JP 3-26. *Counterterrorism.* 13 November 2009.
JP 3-40. *Combating Weapons of Mass Destruction.* 10 June 2009.
Universal Joint Task List (UJTL) [Database with Conditions Version 7.1.] 17 July 2012.
U.S. National Strategy for Counterterrorism. http://www.whitehouse.gov/the-press-office/2011/06/29/fact-sheet-national-strategy-counterterrorism

DEPARTMENT OF ARMY FORMS

DA forms are available on the Army Publishing Directorate Web site (www.apd.army.mil).

DA Form 2028, *Recommended Changes to Publications and Blank Forms.*

Index

Entries are by paragraph number unless page (p.) or pages (pp.) is specified. After a page reference, the subsequent use of paragraph reference is indicated by the paragraph symbol (¶).

A

action element, 7-4
action force, 7-4
action function, 7-1 ¶ 1
actions on objective, 6-27
active supporter, 1-57, **2-41**
adaptive, 2-19
adherent, 6-4, **6-130**
adversary, 1-6, 1-64, 6-8
advisor, 2-29
affiliation, 1-10, 1-57, 2-48, 3-2. 3-13, 4-8, 5-66, **6-124**
air defense, 7-42
all-arms air defense, 7-89
allegiance, 1-56, 2-43, 5-10
ambush, 6-80, **7-11**
 annihilation ambush, 7-11, **7-21**
 containment ambush, 7-11, **7-31**
 harrassment ambush, 7-11, **7-26**
antitank battery, 3-80
antitank platoon, 3-123
anxiety, 6-39, 6-55, 6-75
area of operations (AOR), 3-2
armed combatant, 5-1, **5-10**, 5-90
armor, 2-62, 7-110, 7-181
arms trafficking, 4-73
arson, 4-63, 6-59
assassination, 4-69, 6-83
assault, 7-46
assault element, 7-48
association, 1-57, 2-43, 2-49, 2-61, 3-11, 3-174, 4-30
attack, aerial, 6-49
attack maritime, 6-47
attack on land, 6-45

B

base camp, 3-31–3-37
battle position, 7-124
battle zone, 7-144
bombing, 6-63
breach, 7-77, 7-90
bribery, 4-62

C

cache, 7-91

cadre, 2-36
camera operator, 2-119, 5-93
CBRN, 6-112
cellular structure, 2-72, 2-73
child combatant, 6-91
civil affairs cell, **2-155**
clan, 1-21, 2-43
clandestine, 2-34, 3-28, 6-129
clearing element, 7-6
coalition, 1-67
combat service support, 7-50
combat support, 7-57
command and control, 2-10, 3-1, 4-40
compare and contrast table:
 communications, p.1-17
 composition, p.1-16
 disposition, p. 1-16
 equipment, p.1-16
 finances, p.1-17
 interal security, p. 1-17
 organization-structure, p. 1-16
 personnel, p. 1-16
 recruitment, p. 1-16
 terrorism, p. 1-17
 training, p. 1-17
 uniforms, p. 1-16
 weapons, p. 1-16
complex battle position, 7-122, 7-124, **7-126**
complex terrain, 7-59
computer crime, 4-85
computer fraud, 4-90
computer warfare, **A-44**
corporate organization, 5-25
counterintelligence, 2-41, 2-53, 2-112,
counterreconnaissance, 7-153, 7-175, 7-180
courier, 3-88, 3-145, 4-14, 5-93
criminal activity, 1-73, 4-2
criminal network, 4-30
criminal organization, **4-4**, 4-30
 transnational organization, 4-36
criminal, 1-56, 2-73, **4-1**
field operatives, 4-46

immediate leaders, 4-44
passive supporter, 4-47
senior leaders, 4-41
cult, 1-53
cyber attack, 6-108
cyber crime, 4-81

D

deception element, 7-170
deception force, 7-7
deception, 1-40, 6-39,
defensive action, 3-181, 6-41, **7-122**
demolition, sabotage, and IED team, **2-124**
diaspora, 1-25, 1-60–1-63
direct action cell, 2-11, 2-64, 2-70, 2-71, **2-75** 2-81,
 assassination cell, 2-84
 extortion cell, 2-90
 INFOWAR cell, 2-117, A-15
 kidnapping cell, 2-90
 mortar cell, 2-99
 multifunction cell, 2-81
 rocket cell, 2-99
 sniper cell, 2-84
dispersion, 6-87
disruption element, 7-134
disruption force, 7-2, 7-7
disruption zone, 7-145
drug trade, 4-76

E

economic disruption, 6-137
element, 7-1
enabling element, 7-6
enabling force, 7-6
enabling function, 7-6, 7-75
enemy, 6-5
engineer, 3-97, 7-43
escape and exploitation, 6-29
exploitation force, 7-192
exploitation, 6-29
explosively formed projectile, 6-67
external state, 1-8, 1-64, 5-54
external support, 1-71
extortion, 2-90

F

faith system, 1-52, 2-43

Index

finance, 2-39, 2-90
finance cell, 2-138
fire support, 7-40
fixing element, 7-7
fixing force, 7-7
flexibility 7-8
forces, 7-1
fraud, 4-52
functional tactics, 1-18, 1-72, p.7-1 ¶ 1
fundamentalist, 1-52

G

gambling, 4-57
gang, 4-4, **4-27**
Geneva Conventions, 5-4
governing authority, 1-21
government-in-exile, 2-6, 2-32, 2-37
guerrilla, 1-47–1-48, **3-1**
guerrilla battalion, 3-107
guerrilla brigade, 3-23, 3-46
guerilla company, 3-150
guerrilla company, hunter- killer, 3-163
guerrilla force, 2-2–2-3, **3-1**
guerrilla hunter-killer group, 3-170
guerrilla platoon, 3-159
guerrilla scouts, 3-169
guerrilla squad, 3-160
guerrilla unit, 2-65, 2-72, 3-1

H

hierarchical organization, 1-68
hijacking, 2-90, 4-64, **6-71**
hoax, 6-55
hostage taking, 4-67, **6-78**
higher insurgent organization, **2-48**
human shield, 6-89
human trafficking, 4-74
humanitarian, 2-16, 5-9
hunter-killer, 3-42
Hybrid Threat tactics, 5-188
Hybrid Threat, 1-1, 2-5, 3-10, 6-3, 7-191

I

independent, 1-51, 2-4, 3-2
independent actor, 1-51, 6-4, 6-130
infiltrate, 3-185, 6-6, 7-192
information warfare cell, 2-96, 2-117
information, 1-30
 elements, A-4
 objectives, A-9
 tactical tasks, A-3
 targets, A-9
 tools, A-7
INFOWAR, 1-30, 2-117,

6-104, 7-45, 7-66, 7-92, 7-120, 7-159, 7-188, **A-1**
insider threat, 5-88, **6-99**, A-11
insurgents, 1-48, **2-1**
insurgent irregular OPFOR, 2-1
insurgent organization, 2-1
 higher insurgent organization, **2-48**
 local insurgent organization, 2-13, **2-64**
insurgents-criminals,1-77, 6-76
intelligence cell, 2-105
internal security forces, 1-11, 2-12, 5-32, 6-3
irregular forces, p. viii-ix
irregular OPFOR, 1-1,1-9, p.7-1 ¶ 2
irregular warfare, p. vii ¶ 4,

J

K

kidnapping, 4-65, **6-75**

L

law enforcement agencies, 1-5
legitimacy, 1-7, 5-6, 6-5
local insurgent organization 2-64
logistics, 1-71, 3-21
logistics cell, 2-134

M

main defense element,, 7-132, 7-167
main defense force, 7-5
MANPADS squad, 3-126
media activities, 6-136, **A-40**
media manipulation, 2-96, 2-117, 5-41,6-30
media organization, **5-42**
media personnel, 5-37
medical cell, 2-157
medical section, 3-103
mercenaries, 1-71, 6-3
militia, 3-44, 4-13, 5-31
mortar battery, 3-69
mortar platoon, 3-119
motivations, 1-47, 4-19, 5-5, 5-77–5-78, 6-6, **6-121**, A-33
multile rocket launcher platoon, 3-75, 3-121

N

nationalistic, 1-50, 1-61, 2-149
network organization,2-49,4-12
noncombatant, 5-1
nongovernmental organization, 5-56

non-state, 1-68–1-71

O

objectives, irregular OPFOR 1-12, 1-39
offensive action, 2-180, 3-178, 6-43, **7-9**
OPFOR, p. vii ¶ 1, p. vi ¶ 1
opposing forces, p. vii ¶ 1

P

paramilitary, p. vii ¶ 4 , 1-2, 1-23, 2-43, 3-1, 4-1, 5-2
passive supporter, 2-27, **2-44**
perception management, 1-30, 2-37, 4-110, A-29
planning cell, **2-115**
PMESII-PT, 1-20
 political, 1-21
 military, 1-23
 economic, 1-26
 social, 1-28
 information, 1-30
 infrastructure, 1-32
 physical environment, 1-33–1-34
 time, 1-36
political advisor, 2-33, 2-37, 2-56, 3-57, 3-114,
principles [OPFOR], 1-38
 adaptability, 1-44
 concentration, 1-45
 deception, 1-40
 initiative, 1-39
 mobility, 1-43
 perseverance, 1-46
 protection, 1-42
 surprise, 1-41
private security company, 4-15
private security contractor, **5-13**
protected element, 7-7
protected force, 7-7
psycholgcial techniques, 1-16
psycholgcial warfare, 1-6, 5-21, A-16, **A-33**
public affairs, **A-35**

Q

R

raid, 1-83, 6-80, **7-67**
raiding element, 7-5, 7-70
reconnaissance, 2-205, 3-58, 3-90, 7-37, 7-62, 7-84, 7-108
reconnaissance attack, 7-94
reconnaissance company, 3-90
reconaissance element, 7-7
reconnaissance force, 7-7
reconnaissance platoon, 3-128

recruiting cell, **2-147**
recruitment, 2-14, 6-94
regular forces, 1-22
regular OPFOR, 1-10
regulated military forces, 1-5, 5-4
rehearsal, 6-24
relevant population, 1-6, 1-21 6-145
reserve element, 7-140
rocket launcher battery, 3-73

S

sabotage, 6-61
safe haven, 1-25, 1-34, 1-54, 1-66, 3-21, 6-25, 6-97
safe house, 2-41, 2-141, 3-41, 6-23, 6-29, 7-44
sapper, 3-16, A-12
sapper company, 3-97
security element, 7-7, 7-18, 7-25
security force, 7-7
separtist, 1-50
shelter cell, 2-141
simple battle position, 2-272, 3-183, **7-125**
sniper section, 3-157
special purpose forces, 1-11, 2-3, 3-13, 4-13, 6-6, A-53
stay-behind forces, 3-191

subversion, 1-21, 2-1, 4-3
suicide attack, 6-106
support element, 7-19, 7-50
support force, 7-7
support zone, 7-143, 7-147
supporting cell, 2-18, 2-73
surprise, 1-41, 6-39
surveillance, 2-41, 3-129, 4-50, 5-3, **6-22**
surveillance, pre-attack, 6-22
swarming, 1-45, 3-188

T

tactics and techniques, **7-1**
tailor, vii
target selection, 6-13, 6-20
technical support cell, **2-120**
terror tactics, **6-36**
terrorism, 1-72, **6-1**
terrorism techniques, 6-37
terrorist, p. viii ¶ 3, **6-1**
threat, p. viii ¶ 2
totalitarian, 1-54
toxic industrial chemicals, 2-131, 6-113
toxic industrial material, 2-131, 6-113
trafficking, 4-73
training, 1-60, 2-45, 6-10, A-16
training camp, 3-31–3-37, 3-41
training cell, 2-143
training facilities, 1-66

transnational affiliate, 6-128
transnational corporation, 5-84
transnational network, 1-68
transportation cell, **2-152**
tribe, 1-50

U

unarmed combatant, 5-1, **5-92**
unmanned aerial vehicle (UAV) 3-19, 6-50, 7-113, A-56

V

videographer, 2-79, 5-42, 5-93, 6-30

W

weapons battalion, 3-66
weapons company, 3-116
weapons of mass destruction, 6-54, **6-111**, 6-135
WMD support team, 2-129
woman combatant, 6-91

Y

Z

This page intentionally left blank.

TC 7-100.3
17 January 2014

By order of the Secretary of the Army:

RAYMOND T. ODIERNO
General, United States Army
Chief of Staff

Official:

GERALD B. O'KEEFE
Administrative Assistant to the
Secretary of the Army
1402301

DISTRIBUTION:

Active Army, Army National Guard, and United States Army Reserve: Not to be distributed; electronic media only.

PIN: 103548-000

www.ingramcontent.com/pod-product-compliance
Lightning Source LLC
Chambersburg PA
CBHW071207240526
45470CB00018B/1537